BECOMING A GREAT LEADER AND
BUILDING A WINNING TEAM

THE **Biotech Leader's** Handbook

ANGELOS GEORGAKIS

TO MY WIFE KATYA
for seeing my light when I could not

TO MY MENTOR JOHN MARAGANORE
for allowing me to stand on his shoulders

TABLE OF CONTENTS

INTRODUCTION

The biggest challenge in bringing new breakthrough medicines to patients is not science or money; it's people.
— John Maraganore

This is a best practices handbook for biotech CEOs and senior executives who want to become great leaders and create high-performing teams. This book is about getting fit as a leader, team and company.

The Biotech Leader's Handbook is a collection of short articles and listicles across three leadership pillars:

1. Self-Leadership: Becoming a more self-reflective and resilient leader (self-awareness, resilience, stress management, acceptance, the stoic approach, confidence, assertiveness, self-discipline)

2. Interpersonal and Team Leadership: Building and maintaining strong relationships, communicating effectively, and connecting with others to achieve common goals (relationships, collaboration, trust, communication, management, active listening, emotional intelligence, mastering

1-on-1's, team dynamics, developing empathy, becoming a great coach)

3. Organisational and Industry Leadership: Becoming an inspiring and visionary leader across your leadership team, organisation, and the whole pharmaceutical and healthcare industry (strategy, vision, culture, board and shareholder management, partnerships, business development, industry and thought leadership)

All the tools and frameworks that you'll learn here have been tested and implemented by biotech CEOs and senior executives in my coaching practice.

The Biotech Leader's Handbook is a compilation of writings I have published online, which have been loved and endorsed by respected leaders in the biotech, pharma, investor, and academic community.

The insights you'll find here aren't just mine. I've walked on the shoulders of giants whose names I mention in this book. I'm fortunate to have worked with some of the greatest leaders, mentors, coaches, and teachers. I've learned so much from my clients and everyone in our beautiful community.

Some of you have noticed and asked me why I close all my articles that you can find online with the same phrase: *"Loving you, Angelos."* Love, as leadership, is a conscious practice. Each and every article I've written has been a practice of love towards you, your leadership, your teams, your visions—and my own vision too. I hope you can feel that love while reading this book and pass it on to yourself, your teams, your families, and the patients you set out to serve.

Loving you, Angelos.

LEADERSHIP IS A DELIBERATE AND CONSCIOUS PRACTICE

I've got some good and some bad news for you. The good news is that becoming a great leader and building a winning team is not as hard as getting the biology, the chemistry, the pharmacokinetics, the pharmacodynamics, the safety, and the efficacy of your drug all right.

When you're working on becoming a better leader, you are not walking in a dark room. Leadership development is not rocket science. Getting fit as a leader is the same as getting physically fit in the gym. Fortunately, there are proven strategies and you can come up with a plan that will get you there.

The bad news is that getting *leadership fit*—just as getting physically fit—takes time, attention, and a lot of effort; it's not easy work.

To me, it's absolutely necessary work. I'll go as far to say that focusing all your attention on the molecule and the science and ignoring the health of your company and your leadership is a sufficient condition for failure.

You are in the most difficult business in the world. As the founder of Vertex Pharmaceuticals, Joshua Boger, said, "drug discovery and development is the most difficult activity that humans do."

In this business, there is no room for you to be leadership unfit. Similarly, you can't afford to be unfit as a team and as a company.

So, the leadership wheel has been invented but you still have to spin it hard, a lot, and consistently.

Becoming a great leader is often painful and uncomfortable. It requires ownership, emotional resilience, forgiveness, vulnerability and lots of uncomfortable action.

You are brilliant superstar smart scientists, PhDs, postdocs who have devoted your life to research and studied at the best universities. The edge in this humbling business will not come from having an ounce more of science under your belt; the edge will come from you being a team that works together as one towards the shared vision.

And you being superstars can be a problem in itself! How do you create a superstar team out of a team of superstars? How do you as the leader manage the egos? How do you manage the wet lab vs dry lab fight for glory?

I often say that the biotech journey is a "4x marathon relay race", i.e. early research, preclinical, clinical, and commercial. Different people run the show across the evolution of the company. There are so many transitions in which people pass the relay baton (responsibilities, glory, power) to other people. How do you manage that?

Let's say you are now a clinical company. The current CMO, co-founder, and good friend of yours who has fought for five years to get the company to this point does not seem to have the necessary skills to lead clinical development going forward. You decide to replace them. What do you do with the old CMO? Do you ask them to work under the new CMO? How do you have that conversation? How do you manage the relationship going forward?

To truly innovate in this business, you must be willing to learn by constantly failing. How do you create an environment in which your team is not afraid to fail? Superstars don't like to fail, particularly in front of other superstars. How do you create a culture where your team is not afraid to "break the lab"?

How do you manage your own and your team's emotions when you're reading on Endpoints News that your competitor raised their Series B and has another development candidate while you're still failing your way to success and having only six months of runway?

How do you not get caught up in the short-term noise of analysts, markets, and share prices when running a public company? It's hard! How do you stay focused and play the long-term game you know you have to play? How do you transition from expert to leader and let your scientists run with it? What do you do when you get pulled in all directions by your investors, board, and partners?

These are real issues that are as important, if not more important, than the science itself—and those we'll explore in this book.

The Biotech Leader's Handbook is about becoming a better leader, creating stronger teams, managing relationships with your employees/board/investors/partners, evolving organisational culture and structure, improving communication, but also cultivating self-awareness, being in peace with yourself, growing as a person, and becoming a better human being. At the end of the day, leadership is a self-transformational journey. You build the biotech, and the biotech builds you...

HOW TO READ AND MAKE THE MOST OUT OF THIS BOOK

You don't have to read this book in one go. You don't have to follow any particular order either.

In fact, you must *not* read this book in one go. Why? Because this book is a collection of tools for you not to read and learn, but to practice.

Every bullet point in the listicles is a leadership exercise that will take you weeks, months, until hopefully one day it becomes a habit and part of your company culture.

Read one essay a week or month, and then go and implement. Take massive and conscious action. Put on the new behaviour and see how it fits your leadership style and the team.

Block these practices in your calendar. I ask my leaders to block these practices in their calendar. I ask them, "You say you are a great leader. Show me your calendar." They have blocks in there that say, "Show appreciation to my team, send thank notes to those who exemplified our values this month, journalling, two hours distraction-free deep work, go for a walk outside with my CFO/CMO, email my team about what I am working on with my coach and ask for help, reach out to x and ask to establish a formal mentor-mentee relationship".

IT ALL SOUNDS GOOD, BUT I DON'T HAVE TIME FOR THE LEADERSHIP GYM!

In my work, I can often feel that there is a part of you that wants to shout at me, "Angelos, sorry I don't have time for all these things in my calendar. I'm running like crazy, I'm trying to derisk the science right now, I'm heads down building and pitching to ten investors a day. I don't have time."

I get it. And I'll say, "There is always enough time. If you think you don't have enough time, you are saying, I'm not accepting the rules of this game, which is telling me that you are already playing suboptimally." And we will laugh.

And then I'll say that hitting the leadership gym every day helps you to do all these other things that require your immediate and full attention more productively—and for longer. You go to the leadership gym to scale yourself and your team.

At the end of the day, in the most high-risk high-failure business of drug development, you go to the leadership gym to build a good company and feel good about your team and the work you do every single day.

This business requires a lot of humility. You can't control biology, you can't control the science.

But you can control the company culture, you can control the experience of your team working for this company every single day. No disappointing data, no clinical trial failure, no gloomy fundraising or economic environment can take that away from you.

That's your hedge against failure. You will fail, you will pivot, you will start all over again. To keep going at the science gym, you need the leadership gym!

WHY I DO WHAT I DO & A PERSONAL NOTE

This is the graph that I have on the corkboard in front of my desk (see page 8).

I am this little dot on the left. If I can empower my leaders who can then empower their teams who can then do great work that reaches the patient, then, I can reach the patient myself, and have my own impact on human health.

That's my why—and the message with which I close all my writings that you can find out there.

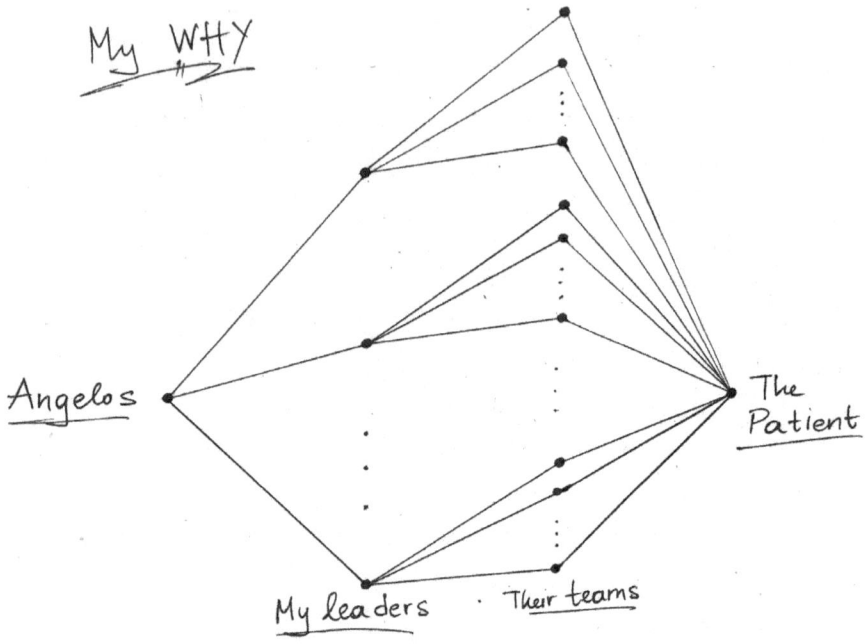

I am on Zoom or the phone all day long talking to biotech CEOs and executives. Our conversations are creations. Sloweddown, soul-to-soul conversations. This work is for me the highest form of creation.

I love my leaders enough to be whoever I need to be and do whatever I need to do so that they may see.

And because my leaders can feel that love, they will allow me to often challenge them and go to uncomfortable places in my commitment to truly support and serve them.

This book is full of advice. However, my role is often just to listen and say nothing. The CEO's job can often be a lonely job. My CEOs share with me their deepest fears and anxieties, the things that they can't share with their boards, investors, teams, and even with their life partners.

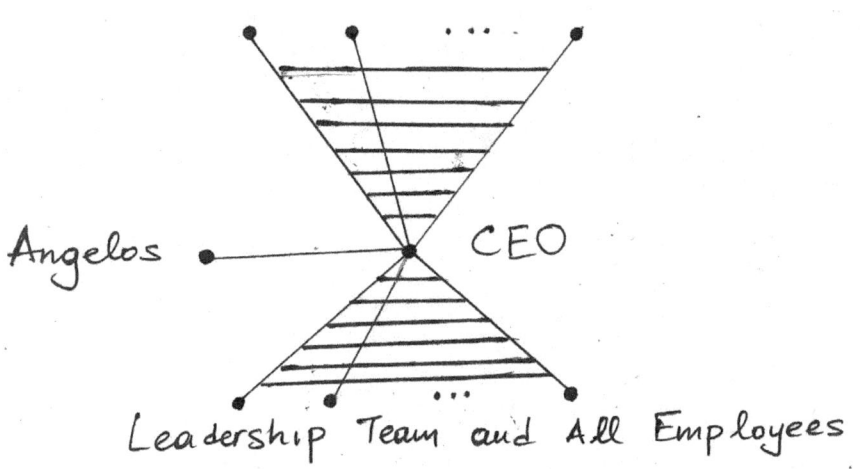

My role is to help my CEOs get clarity on what they have to do, trust their intuition, and come up with their own customised recipe that often doesn't exist anywhere out there—including in this book. So what we do in our work is the tailoring!

You can find more about my work on my website biotechsuccess.com. Write to me, say hello. I'd love to meet you and get to know you.

Loving you, Angelos.

TEAMWORK MAKES THE BIOTECH DREAM WORK

1. HIRE "POLYGLOTS".

"It's hard to build a multidisciplinary team from hyper-disciplinarians. People who speak multiple languages have an easier time learning another one." — George Church

"Bilinguals can bridge the chasm and play the role of the translator between disciplines. They are rare but having those at the very beginning is extremely valuable. Over time, focus on growing bilinguals internally." — Daphne Koller, CEO at Insitro

Sri Kosuri and the team at Octant Bio call themselves "ante-disciplinary", i.e. they hire generalists who are working across domains. *"It has downsides but it's the only way to tackle hard problems and chart new approaches in drug discovery."* — Sri Kosuri, CEO at Octant Bio

2. PRESS HARD ON THOSE REFERENCE CALLS.

This is something that first-time founders and CEOs like to skip. There's no way you can hire Lucy unless you've interrogated some people that Lucy has worked with in the past.

So, you don't just ask, "is Lucy a team player?" You want to really dig in:

- Tell me about a time when you and Lucy worked well together.
- Tell me about how Lucy handled a disagreement or conflict.
- How did Lucy change after receiving feedback from you or the team?
- How did Lucy go above and beyond to support the team?

3. TEST THEIR COMMITMENT TO THE VISION.

Don't whitewash the risks when hiring. Tell them they're singing up for the scariest roller coaster ever. But, if you succeed, you'll redefine human health. You want the right people to select in.

Hiring people who are bought into the vision is a first-level immunisation against bad culture. Because when people are willing to die for a cause, they are more likely to set aside their ego for the benefit of the team when conflicts arise.

4. PAY ATTENTION TO THE INTERFACES.

You have all these amazing people who are experts in their fields. The problem—and your biggest challenge—is that these people speak their own language and do things their own way.

You have to build a translation layer and formalise the way that information moves between domains.

Everyone in the team must be clear on:

- where their work is
- how their work impacts the next person's work
- what are the requirements of the next person and why

From Day #1 the founder/CEO has to facilitate the dialogue between domains to build those cross-cultural bridges and make sure people communicate well with each other.

5. INSTILL A "NO-LANES" MINDSET.

Although everyone's responsible for their own area of expertise, there's no such thing as "you stay in your bio lane, you stay in your data lane". Everyone has to think across the entire product and understand how all the pieces fit together. Encourage everyone to challenge assumptions.

Hani Goodarzi, Associate Professor at UCSF, says: "I don't think in wet or dry terms anymore, I just think in terms of problems and solutions. It's always a problem that we as a team are trying to solve."

The onboarding process is critical. Create a slide deck that gives a complete picture of the whole system end to end, with all the different disciplines and interfaces. Have new employees talk to all teams.

6. EVERYONE NEEDS TO HAVE A FORWARD-LOOKING APPROACH.

In drug development, there's nothing that a person or team can do without affecting another person or team. Often, best practices in one team may create problems in another team which means that a team may have to compromise its standards for the common good.

Remember the classic from Systems Theory: "to optimise the system, you have to sub-optimise the sub-systems."

Say to your people, "You are not a data scientist, you are not a computational biologist, you are not a research associate. Your job is not your job. Your job is to help the team win." The ultimate purpose of whatever you do is to help someone else in the team, and by induction, to help the whole team win.

7. EVERYONE SHOULD HAVE THE END GOAL IN MIND.

It's very easy for a team to get so immersed in what they're doing and lose the forest for the trees. The goal is not to complete a trial or discover something cool.

Everyone should do their work with the end goal in mind: to develop a drug that can be registered and that addresses an unmet medical need. That's why it's crucial to bring in commercial input as early as possible.

The forward-looking approach applies to partners, pharma, and regulators too. There's always a recipient of your work either inside or outside the company. The key is to be mindful of the recipient's requirements right from the start and throughout the process.

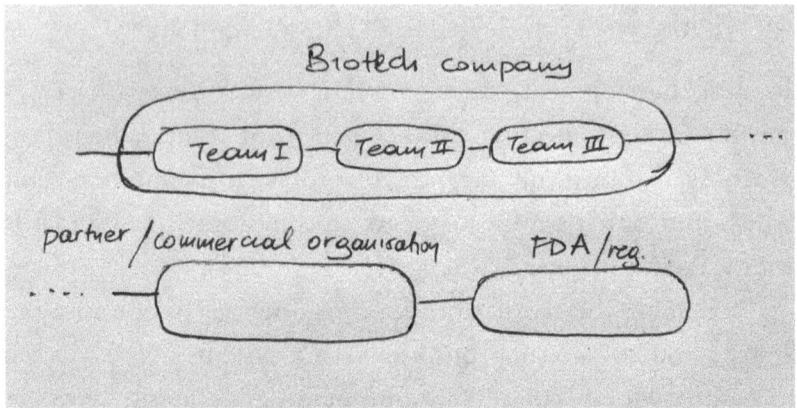

8. BE OPEN AND HONEST ABOUT WHAT YOU DON'T KNOW.

Humility—like all virtues—is modelled at the top. If the leaders are open about their blind spots, the team will be too. If the leaders won't admit their gaps and weaknesses, the team won't either.

Talk openly in all-hands meetings: "These are the areas I'm working on and I need your help". Send your performance reviews from the board to everyone in the team.

9. ENCOURAGE OPEN EXPRESSION OF PERSPECTIVES AND IDEAS.

Make an explicit rule for this principle. Talk about it—and walk the talk. Again, it starts with the CEO and executive team.

When someone asks a "stupid" question or comes up with a "naive" suggestion: "Why don't we do x?" Instead of saying, "because this is not how it's done", have everyone think about it: "really... why don't we do x?" This can lead not only to better answers but better questions.

10. DON'T STICK IT IN THEIR EAR.

You've hired these brilliant people, now trust them with the freedom to execute. You don't tell smart people what to do. Smart people don't respond well to that. What you do is, hopefully, you inspire them to want to take action.

11. TELL THEM HOW AMAZING THEY ARE; KNOWING IS NOT ENOUGH.

Everyone is an expert in their field and wants to be acknowledged and celebrated. Add a weekly "love the team" block in your calendar to show them your appreciation. One by one, not a "hey everyone" in Slack!

When you catch yourself feeling grateful about someone or something they've done, let them know. When you hear something nice said about someone, let them know. And be specific: "Jen, I appreciate you for updating those process documents".

12. FILL THE GAPS BETWEEN PEOPLE.

Pay attention to body language, mood shifts, and side conversations. Spot when someone is frustrated. Listen and watch intently to correct miscommunication. Use your 1-on-1's to keep a pulse on how people are feeling and to uncover tension.

13. GIVE CREDIT TO EVERYONE AND HELP THEM SHARE THE GLORY.

Your goal is to be a good parent to all your children: lab scientists, data scientists, engineers, etc. For example, lab scientists may see the help offered by data scientists as "threatening to their glory".

Address and discuss these concerns with the teams. For example, a solution could be that the wet lab doesn't completely hand it off to the data team. They may decide to do the analysis and/or the presentation together. Make sure everyone is in the room and gets credit for their work when results are presented. Again, thank people one by one in those meetings.

14. BE TRANSPARENT ABOUT CHANGE.

In a biotech company, things change so rapidly: priorities, targets, strategy... There are so many transition points: from preclinical to clinical, from private to public, from clinical to commercial.

You have to be transparent about these transitions if you want to retain your incredibly talented people and attract new talent when you need them. Everyone has to be clear on where they stand and how their circumstances might change.

The best book I've found on this topic is The Hard Thing About Hard Things by Ben Horowitz. There is no playbook for being a CEO, but what this book does is it helps you see the company and your decisions through everyone else's eyes.

15. BUILDING A GREAT COMPANY CULTURE IS NOT ROCKET SCIENCE. HOWEVER...

Founders/CEOs see culture as an elusive thing because simply they're not intentional about it. Building culture is easier than designing the right molecule for the target. However, if all your attention is on the molecule and you ignore the health of your company, there's no way you can get the molecule right. Bad culture will break the company. Team building and company culture require as much attention as science does. I call it "the leadership gym". Your muscles and body need exercise to be strong; the same applies to culture and leadership. It takes attention, it takes time, it takes reps.

BENCH SCIENTIST TO BIOTECH CEO IS A BRUTAL TRANSITION— SIX KEY MINDSET SHIFTS FOR THE BIOTECH LEADERS OF TOMORROW

1. FROM SCIENTIST TO POLITICIAN

The biotech industry has some unique challenges:

- long lifecycles
- capital intensity
- high risk
- pharma alliances
- regulatory scrutiny

To navigate these challenges you need to build strong communication, sales and networking skills. You have to be incrementally better at fundraising and attuned to corporate development. You can't be an isolationist building your product and not worrying about anything else. You should be developing relationships with other companies, investors, and partners from Day #1.

2. FROM SOLO EFFORT TO TEAM EFFORT.

Building a biotech is a professional team sport whereas academia can often be isolating. It's about "how we can attract the right people to get this done as quickly as possible" vs "how I can do this with the least amount of help."

An orchestration of minds is required to reach the patient. Both within the scope of the company with a multidisciplinary team, and outside the scope of the company with partners, pharma, and regulators.

Bob Langer has said, "you don't want to feel threatened. You must hire people who are better than you."

I say, "you must feel terrorised! If you're not feeling intimidated by the brilliance of people you're bringing on, you're doing it wrong!"

To reach the patient there's only one way: Together. So...

- Identify your weak sides.
- Ask for help.
- Seek advisors and mentors.
- Communicate and be open.

- Pressure test your assumptions.
- Hire people who are better than you.
- Give your team freedom.

3. FROM SCIENTISTIC TO MANAGER.

Your role will soon be to serve your team and ensure they have everything they need to succeed. Your output is no longer *your* output. Your output is now *their* output. It's what *they* can accomplish vs what *you* can accomplish.

Management and leadership are the biggest areas for growth among scientific founders. Where should you start? Read the book High Output Management by Andy Grove.

4. FROM YESES TO NOS.

"More people have said no to me in the last 18 months than in the rest of my life combined." — Adam Freund, CEO at Arda Therapeutics

You'll get a lot of nos. But nos are not a reflection of your merit. They often have nothing to do with you.

For example, take fundraising. There are 100s of reasons a VC may say no to you that you have little to zero control over:

- bad memories of deals gone south
- internal politics
- conflicts of interest with other portfolio companies
- the VC's own unconscious fears and biases

What's your best shot? Try to understand the person in front of you. Deeply.

Ask direct questions: I understand our approach may sound complex. Is there anything that's not clear to you? What are your concerns?

Remember that you're an expert in your field. Investors won't sit down to argue with your science. They know less than you. No one has had the insights you've had. If you want people to trust you, you have to trust yourself first. Persevere. Keep building. They'll follow.

5. FROM ENEMIES TO ALLIES:

If you think that investors, partners and regulators exist to make your life harder, you won't get far. You'll have to work together with these people. You need them as much as they need you. Partnerships require trust.

- Reach out to investors and partners early on.
- Give information and don't worry about IP protection or your ideas being stolen.
- Ask questions, listen, and learn.
- Understand their goals and concerns.
- Find how you can help them so they can help you.

6. FROM PERFECTION TO ACTION.

Founders often see an encounter with investors or pharma as a one-shot game, i.e. you arrive in a meeting with your perfect data and they either say yes or no. No, no, no. These are your clients, your market. They are the ones who will tell you what data they need.

Build backwards:

1. First, talk to people in the industry, investors, pharma, and other companies to understand the marketplace.
2. After you've confirmed there's a strong market need, generate data and build.
3. Show them your progress and request feedback.
4. Repeat.

Having a closed and frequent feedback loop where you pressure test your solution against investors, pharma and partners leads to a better product and faster development cycles.

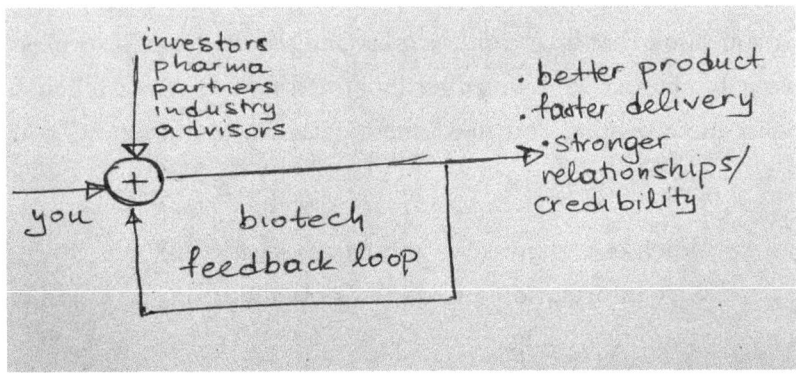

The more you ask for their input, the more they like you. The more they see you grind, the more they trust you. This is a relationship game. At the end of the day, people do deals with those who they like and enjoy working with...

10 FUNDRAISING TIPS FOR BIOTECH FOUNDERS/CEOS

1. BE AN FBI INTERROGATOR.

I often hear, "I've spoken to 50 investors. They said they want to see more data/proof". And I go, "What data/proof do they want to see specifically? Did you ask them? Did they tell you?"

You have to really push for answers here. "I need to see more data/proof" means absolutely nothing! When they say something like that, interrogate them like hell! And I don't mean confront them; I mean genuinely ask questions to understand what they mean.

"What is it exactly that you'd like to see? What data? What progress?"

Be open and direct, "Listen, how do I make you go WOW next time I reach out to tell you about our progress?"

"I'm sure you've spoken to people who're building in the same area. What—if accomplished—would make us really stand out from everybody else?"

"What do you want us to derisk next?"

Don't hesitate to ask these questions. The quality of your questions, the straightforward approach, and your perseverance... are all things investors want to see in you.

Then, investors will give you some "first-level" information. At that point, you have to "coach them" further, i.e. help them think through things to see more clearly what would make a difference and why. It may not be clear to them right away. Ask good questions.

"Why do you think this is important? Why would this increase your confidence in our technology?"

"Let's suppose we've proved that. Why are things now easier? What comes next?"

Help them to slow down and reflect. Fill in the gaps for them.

When they're finished, don't rush to say "thank you". Playback whatever they said and confirm understanding:

"Okay, let's see if I got this right. What I'm hearing you say is that you want to see x, y, z because... is that correct?"

"Yes, that's right". Mission accomplished. You got what you needed.

Now imagine the trust you build with an investor when you go back to them after 3 months and say, "Hey, here's the progress you wanted to see." (Well, then they'll ask for something else but that's fine!)

Now... you don't do your homework or ask questions to make investors feel important or please them. Investors are closer to the market than you, they do due diligence on hundreds of companies like yours, they speak to pharma... Your job is to "steal" their market intelligence.

Oh, and the same applies to pharma, i.e. your buyers. They'll give you their shopping list, they'll tell you what they need from you, where they've struggled and how you can help.

Your job is to collect market insight as early as possible from both investors and pharma.

This approach has multiple benefits:

- you build the relationships (biotech is a relationship business)
- you pressure test your assumptions
- you learn
- you derisk the market

2. PERCEPTION IS REALITY (IT APPLIES TO INVESTORS TOO).

Investors love $$$—and their job is to pick the winners who will generate the biggest returns for their limited partners—but they're into this (especially in life sciences) because they also want to have an impact.

Is this true for every investor? No. This is simply what "I choose to believe".

And I do so because, first, I've met people like that, and second, this is what serves me to believe.

Because, by proof of contradiction, how can the belief "there are no good investors out there who are willing to help me" help you and your company?

In the end, how can a belief like that help your state of being when talking to investors?

On the contrary, when you keep looking for the good investors out there because you believe those exist, you'll definitely see more of them. It works with investors; it works with everything in life!

3. DON'T TALK TO THE INVESTOR; TALK TO THE HUMAN BEING.

Investors are human beings. Tell them, "Listen, I care dearly about this bloody thing. I want to get this technology to the patient and save lives...

"If for whatever reason you're not ready to invest, I understand. All I want is your honest opinion, an insight, something you happen to see more clearly to which I'm blind to. Please tell me what you honestly think so I can get one step closer to my vision".

People want to help (another thing that's up to you to believe).

Most investors will pass, but why wouldn't they do something for you, if you genuinely asked? And the least they can do for you is give you their honest opinion.

When I fell in love with biotech and decided to work only with biotech leaders, I was amazed at how tiny the community is. Everyone knows each other. So, why wouldn't an investor be nice and helpful to the community?

4. TALK TO THE SCIENTIST IN THEM.

In the life sciences, most of the investors are scientists, PhDs, and postdocs like you! At some point in time, you guys could have been working together in the same lab, on the same thing!

Talk to them scientist to scientist. Speak to what unites both of you. You can even joke with them, "Sometimes I wish I was on the other side!" :)

5. PLAY THE STEVE JOBS CARD.

When things get tough, say what Steve Jobs said to Sequoia's Don Valentine, "Tell me what I have to do to have you finance me". Don Valentine passed but, to "soften the blow", he introduced Jobs to the people who eventually... funded him! Another proof here that people want to help...

Can you be so passionate about your vision that you give investors no alternative but to want to help and do something for you?

Steve Jobs' Fundraising Marathon

angelosgeorgakis.com

Key

----→	Rejected
----→	Invested
──→	Introduced

6. FUNDRAISING IS NOT A ONE-SHOT GAME.

Fundraising is a process in which two parties get to know each other, build a relationship, and develop trust before they decide to work together. Trust takes time and a lot of "touchpoints".

7. FOLLOW UP ALWAYS WITH UPDATES.

Don't just say: "Hey, did you get my email?" Include as many updates and wins as possible. Say, "George Church joined our advisory board! We got another $200K grant. We hired a computational expert."

And if they don't always reply, don't stop. Fire that email to everyone on your list. It doesn't cost you anything...

8. IF YOU WANT MONEY, ASK FOR ADVICE. IF YOU WANT ADVICE, ASK FOR MONEY.

This is particularly true in the current fundraising environment (May 2023). Tell them that you're not ready to raise yet but you want to build a relationship with them and also ask for their opinion. what's a better time to build relationships with investors than now?

9. PITCH ONLY TO SECURE THE NEXT MEETING.

In your first pitch aim for "clear and convincing", not "detailed and comprehensive". Give them a "hook", a shorthand way to remember and describe your company. The VC who heard your pitch now will need to convince their colleagues in the VC firm that you deserve a second meeting. And they'll probably have sixty seconds to pitch your company in their team meeting.

What do you want them to remember? You have to spend a great deal of time thinking about that take-home message. And that message should be summarised in bullet points both at the beginning and the end of your presentation.

10. PITCH LIKE ECKHART TOLLE.

This is probably the best advice I can give you on how to stay motivated during the draining fundraising process. I also call it "conscious" or "mindful pitching".

As I said earlier, VCs are human beings. And as all human beings, VCs are governed by fear, they get stuck in doubt, they're stressed, they get greedy, they become emotional, they have limiting beliefs, they make irrational decisions.. They're anxious about delivering results, they get bored, they're impatient… And they talk to 15 entrepreneurs a day looking for a needle in a haystack. For all these reasons, don't just "fire" your pitch.

Remain conscious. Connect with the human being behind the VC.

You've scheduled 30 minutes to talk to this person; 30min for you, and 30 min for them. Enjoy these 30 minutes. Be candid, open up, share, and build a relationship with them.

A relationship that can benefit both of you today, in six months, or five years. It's not just a yes or a no. It's the beginning of a new relationship. Give yourself and them the gift of slowing down and remaining present in your conversation together.

This person may be your future partner. Enjoy the person in front of you. It's all you have right now. Enjoy that interaction even if it's the last time you'll see each other.

VCs will not invest in you because of your spreadsheets. They throw darts at the wall anyway.

They'll invest in you because you were the only one from the hundreds they've talked to who taught them Eckhart Tolle's Power of NOW! They'll invest in you because this chat felt different from all the other chats.

And they may have no clue why it did so! They are often unconscious—and unaware of their decision-making process. They're going at 100m/h just like you. Slow down both yourself and them—and win. This is leadership; lead yourself into conscious pitching so that you can lead them into conscious listening, and hopefully, conscious investing into your company!

10 FUNDRAISING TIPS FOR BIOTECH FOUNDERS/CEOS SUMMARY

1. Be an FBI Interrogator. Push investors for honest feedback. What do they mean when they say, "I need to see more data/proof"? If they're not ready to invest yet, what progress do they want to see before they do so?
2. Perception is reality [mindset]. What you choose to believe about investors will manifest in your relationships with them. "When you look for something, you see more of it." If you believe there are good investors out there who care and want to help you, you'll see more of those.
3. Don't talk to the investor; talk to the human being. Human beings are wired to help other human beings. Tell them how much you care and genuinely ask for help.
4. Talk to the scientist in them. These guys are not just VCs. In the life sciences, VCs are usually brilliant

scientists like you. They've done their PhDs/postdocs. Speak to the things that unite you.

5. Play the Steve Jobs card. "Tell me what I have to do to have you finance me". Be direct, persevere.

6. Fundraising is not a one-shot game. Fundraising is about the relationship, the trust, and VCs seeing your progress/evolution as a leader and company.

7. Follow up always with updates. Don't just say "hey, did you get my email?". Include updates and wins in your followups. What have you achieved since the last time you wrote to this person?

8. If you want money, ask for advice; if you want advice, ask for money! Investors may not be able to invest in you right now (especially during the current environment) but they can be generous with they advice and insight. Reach out today, focus on the relationship, learn, and ask for advice.

9. Pitch only to secure the next meeting. In your first meeting, aim for "clear and convincing", not "detailed and comprehensive". Your primary goal is to help the investor see the opportunity and be able to "pitch" that opportunity in 60 sec when they meet with everybody else in the VC firm the following Monday...

10. Pitch like Eckhart Tolle (The Conscious/Mindful Pitching). Remain conscious, be present. You've scheduled 30min to talk to this investor, 30 min from your life and their life. Connect with the human being in front of you. Enjoy this conversation.

BIOTECH IS A TEAM SPORT

When George Yancopoulos, the Cofounder and CSO of Regeneron, was a postdoc at Columbia and received a faculty position offer, the Department Chair took him out to dinner and said:

"I want to be clear with you about one thing. Once you start your own lab, you can't publish with Fred anymore."

(Dr. Fred Alt was Yancopoulos' supervisor, and mentor at Columbia for ten years.)

"Why?", Yancopoulos asked. "The whole reason I want to be here is because I want to continue publishing with Fred!"

"You have to prove that you are independent now; you have to show us that you can do this without Fred."

Yancopoulos responded, "Wait... let me get this straight. If I can cure cancer with Fred Alt, that's not good enough; I have to cure cancer on my own?...

... how about Brown and Goldstein? Why can't I and Fred be like Brown and Goldstein?"

(Brown and Goldstein, who won the Nobel Prize together in 1985, were Yancopoulos' heroes at the time... and they've been on Regeneron's board for more than 30 years!)

"You are neither Brown nor Goldstein", replied the Chair!

Yancopoulos later reflected, "I understood that I would have to do it alone without Fred, which not only took away the fun but also made Fred my competitor!

Fortunately, it was around that time when I received a cold call from Len Schleifer, and together we founded Regeneron.

At Regeneron, I wanted it to be like the early days with Fred when we were all one family, working together and supporting each other.

That became our model. It's what we said to the first people we hired, and it remains our model today. Here at Regeneron, no one works in isolation..."

The only way to build an iconic biotech company and bring breakthrough medicines to patients is..."together".

And that's the biggest mindset shift for you guys coming out of academia.

So...

- identify your blind spots and find people who complement them early on
- seek mentors who've done it before
- build a strong and active Scientific Advisory Board and Board of Directors
- pressure test your assumptions
- hire people who are more skilled or experienced than you
- hire your "counterbalances", i.e. people who have the same vision but a different approach, mindset, and perspective on the world
- surround yourself with people who challenge you
- choose investors who're going to fight in the trenches with you for years...

ps: "Back in the '80s, biotech was like going to the dark side, it meant you were lost forever. It was the time when the first genes were cloned and Schleifer was desperately looking for a gene cloner. It turned out I was the only one who answered his calls!" — George Yancopoulos

DANIEL SKOVRONSKY'S
LEADERSHIP PRINCIPLES

Daniel Skovronsky made Eli Lilly the most valuable drug company by market cap (Sep 2023). His company was acquired by Lilly and within 8 years he became Lilly's CSO. He took brave bets in areas no one was willing to go for patients—and he won.

Here are eight principles from Dan Skovronsky's leadership playbook:

1. WORK ON SOMETHING REALLY HARD FOR A LONG TIME.

Skovronsky was at home when he got a call from Lilly's statisticians informing him that their drug was effective. Donanemab could slow the progression of Alzheimer's by 35%.

"Any words I use to describe my emotions at seeing the data will fall short. I've been pursuing the same enemy for 25 years. I was a kid when I started at this. Working on something hard for a long time and seeing it come to some meaningful fruition is hugely rewarding".

"Now I want to make sure that my employees have this kind of experience: Work on really hard problems for a long time, slowly building up the scientific evidence until we can make meaningful drugs, and then, do bold trials and see the results."

2. WANT TO CREATE CHANGE? BE BOLD AND CRITICAL.

"One reason he succeeded, he believes, is that he was willing to be critical because he already had success. (Skovronsky joined Lilly in 2010 when Lilly bought his company Avid Radiopharmaceuticals for $300m.)

That made him bold, he said, and led him to point out how slowly Lilly moved in developing new medicines. Lilly executives responded by offering him a role as the company's vice president of target therapeutics.

Over time, his complaints would have an impact: Since 2014, the time it takes for new medicines at Lilly to go from starting clinical trials to approval went from 11.5 years to 6. Within two years, he was promoted to SVP for product development and within eight years to CSO."

3. MARRY THE BEST OF BOTH WORLDS, BIOTECH AND PHARMA.

What's incredible about pharma is the resources, the people, the money—and the stability and continuity of that—which leads to institutional learning, investing in functional know-how, and the development of expertise.

What's incredible about biotech is the conviction, a sink-or-swim around a conviction, an asset, a technology or an idea. The focus that comes with that conviction is hugely valuable. The inadequate resources force you to focus on the best opportunities.

Because you have no stability in biotech, you sweat the small and the big stuff. There's personal involvement from the leaders because their lives are at stake. And that's why everyone is involved in the details of clinical trials, regulatory affairs, discovery, strategy, etc.

What I've tried to do at Lilly—and I'm still not fully successful—is to marry the best of both worlds: within the walls of a large pharmaceutical company, with all the resources, create smaller groups that look like biotechs.

We have companies that we have acquired over the years—including the company that I came from 10 years ago—that still exist as biotech companies within our walls operating in the same way as they did before.

The biggest of those companies is Loxo Oncology. What we did with Loxo was, not only did we bring them in and say, keep doing what you're doing, but we gave them the rest of our oncology portfolio and said, now run our portfolio the same way!

What I've tried to do over the years is to abolish as many committees as possible and dismantle the bureaucracy that pharma companies usually rely on. We set up the molecules in our portfolio as little companies and each one of those companies has a leadership team and board of directors. And they'll meet whenever they need to—and that's the way we provide the governance instead of the teams coming to committees for approvals.

The teams love it, people find it so satisfying, and decision-making is faster. The leaders of the various functions/committees—who you might guess would be opposed—tell me, "I love being on the boards of these companies within Lilly, I'm engaged and close to the projects, it's fun!"

4. WHAT MOTIVATES PEOPLE IS NOT THE REWARDS BUT THEIR PERSONAL IMPACT ON THE WORK.

"When we think about biotech, people usually jump to the asymmetry and the outsized financial rewards for risk-taking, but I haven't found that to be the real problem.

It's all about being able to see your personal impact on the work—and that's the valuable thing about biotech that you have to create in a big pharma company. People have to see how their work matters, how it impacts the project, how they're making a difference."

5. DELEGATION BECOMES EASY WHEN YOU HAVE THE RIGHT PEOPLE.

"Giving up power and delegating is not easy; we all struggle with it. That's why people are the most important ingredient. If you have great people whom you can trust, why would you be uncomfortable delegating to them?

Delegation also becomes easier when you have acquired a biotech company. We paid billions of dollars because we like the decisions these guys have made over the last few years... Well, let them continue to make those decisions and figure it out themselves!"

Whenever people tell me about their best career experiences at Lilly or anywhere, they always say, "I was working on a small team, there was time pressure, it was really important for the company, and it was delegated for us to figure it out—and we did it!"

6. PERSEVERE WHEN EVERYONE ELSE GIVES UP.

In 2016, Lilly's lead Alzheimer's drug, solanezumab, failed dramatically and caused a multi-billion-dollar plunge in Lilly's market value. That was a profound blow to Skovrosky's hypothesis that targeting amyloid could ever lead to an effective drug.

Major drug companies such as AstraZeneca, Pfizer, BSM, GSK, and Amgen, were giving up on brain drugs. Pfizer closed its neuroscience division and cut 300 jobs.

"Dave Ricks wasn't yet our CEO at the time but he was running the business unit that controlled that asset. So Dave asked me to do a special project, which was to write up a rationale to show to our board of directors what we would do next if solanezumab failed.

I worked really hard on that, and as a scientist, I dug deep into all the reasons from genetics to mouse models to pathology and biomarkers, why this is the right target, and why we should continue going after it with better and better drugs.

When I showed it to Dave, he said, "This is great, except you forgot the most important reason why we're not giving up on Alzheimer's. And I was like, oh no, what is it?

The most important reason, Dave said, is the patients who are suffering. And if we don't do it, no one else will. He was right...

That's what it takes in this business. You've got to combine a passion for helping patients with deep science. Alzheimer's is the most feared disease among elderly Americans and the only of the top causes of death that we've had no progress over decades, so we couldn't give up.

Sometimes disease areas become popular, and sometimes they fall out of fashion. For the last ten years, Alzheimer's was way out of fashion. The strategic thing to do was to stop working in neuroscience. That shows what a great leader you are. We resisted that."

7. PERSEVERE BUT LEARN FROM YOUR MISTAKES AND MAKE SMART DECISIONS.

Picking up the pieces, Skovronsky and the team would target only what could be reliably measured in the brain going forward. Drugs would also have to hit their targets hard to advance to later stage development.

They also took another strategic decision to focus on Phase 2 trials rather than rush from early studies to giant Phase 3 trials with thousands of patients as everyone did before in the hope of winning the Alzheimer's race. Well, at that point, they had no competition anyway!

The fruit of seven more years of perseverance and learning was donanemab—a drug that showed a 35% slowing of Alzheimer's progress that's now on its way to FDA approval.

8. BE TRANSPARENT AND THOROUGH—ESPECIALLY WHEN THE COMMUNITY IS SCEPTICAL.

Matthew Herper at STAT News pointed out at the STAT Breakthrough Summit in May 2023 that Skovronsky had put a lot of data in the press release.

Skovronsky said, "There's so much scepticism in Alzheimer's because of the history of failure. We didn't want people to look at the data and say, "Well, they didn't tell us the secondary or maybe it was just a single analysis method, or we didn't exactly see the p-values. We didn't want more controversy in the field. I don't think that's helpful for science, and certainly, not helpful for patients."

THREE MORE THINGS:

1. There's something else that Dan Skovronksy said during an interview with Mike Rea IDEA Pharma that I found to be thought-provoking and at the same time very humbling:

"We are at this point where we have unlocked the science and we've translated that into some powerful medicines. Now... making those medicines into widely available global solutions for hugely important diseases is equally if not more challenging. Probably earlier in my career, I wouldn't have recognized the challenge of that half of the business.Just creating a medicine and even getting it approved isn't enough; you've got to get it available and accessible to people and make sure it's used. Developing better drugs is hard but creating a brand new standard of care turns out to be even harder.One of my personal ambitions is that we see more of the innovation culture applied to the commercial side of our business."

2. Dan Skovronsky wouldn't have been able to create change and innovation had it not been for the support of Lilly's CEO David Ricks. Ricks backed Skovronsky in a way that is rare in the industry.

"Dan's different. I had this sense immediately. Some people get in biotech for the cha-ching..." — *Dave Ricks.*

3. The point above reminds me of what my good mentor John Maraganore has said, "One person cannot do it alone. Leaders need someone else to help them complete the sentence, close the door they opened, or see around the corner."

MENE PANGALOS' LEADERSHIP PRINCIPLES

He 5X'ed AstraZeneca's productivity from pre-clinical to phase III completion and transformed the company's R&D strategy, culture and approach to innovation.

The more I researched Mene Pangalos and immersed myself in his interviews, the more I found myself inspired by his exceptional communication skills and his ability to unite people around a common vision.

10 key principles from Mene's leadership playbook:

1. WE DON'T DO BACKUPS ANYMORE!

"When I joined in 2010 I tried to get everyone bought into reasons why we needed to change and learn from what we had done before. So, we looked at all of the projects that were run from 2005-2010. We were spending about $5B a year on R&D.

When we did the analysis, we found that if you measured us by the number of things that we were doing, i.e. the number of candidates that we were putting into the clinic or the number of IND's that we were filing, we were the second most productive company in the industry. But if you measured us by the number of launches, we were the industry's second least productive company.

There was a disconnect there. Our science was getting rewarded, but no medicines were coming out at the other end.

For example, we used to have so many backups in the pipeline... Backup #1,2,3,4,5,6... and then of course all those backups had exactly the same probability as the lead molecule!

It's the quality of what you work on, not the quantity of what you do. We don't do backups anymore."

2. THE 5R FRAMEWORK.

"Based on the data that we analysed, we found five things that would improve the probability of running a successful programme.

The 5Rs may seem pretty obvious, pretty intuitive, yet actually quite difficult to execute on consistently."

Right Target

• How well do you understand the biology of the target?

- How well do you understand the disease pathophysiology, how it connects/relates to path you're trying to modulate?
- What genetic validation do you have in pre-clinical animal models or in human genetics?

Right Tissue

When you have a molecule, whether it's a monoclonal antibody, a small molecule, or whatever the drug modality, demonstrate first of all in the preclinical models that you can engage the target and understand your PK/PD relationships.

"If you can't demonstrate target engagement in a clinic, we have a big problem, because then if you fail you have no idea if your molecule is crap or if your hypothesis is wrong. So, a good failure is for me is one where I know have demonstrated target engagement but the molecule didn't work, so biology is wrong.

We were running a number of Phase II where the molecules failed and you asked the question, 'Okay, so it didn't work, did we engage the target? Did the receptor antagonist get into the brain if it's a schizophrenia program, and what you got was quizzical blank stares from everybody saying, we have no idea. We were not learning anything because we had no idea why we were failing. That doesn't happen anymore." — Mene Pangalos

Right Safety

Establish safety as far as possible in humanised systems before initiating clinical trials.

"Because our scientists were being rewarded for the number of candidates, they were remarkably good at lowering the doses to the minimum amount, where they now—because they're not measuring target engagement—engage the target but they still get the candidate through.

And what we saw was that when you had early safety signals, they invariably came back to bite you somewhere during early development, or even worse, later stage development. So, waiting out your safety signals early, making sure you are working on the right series, on the right scaffolds, that you understand both your target-based toxicity and your molecule-based toxicity is really important." — Mene Pangalos

Right Patient

Find the patient population in which your medicine is most likely to work, because if It doesn't work in that patient population, it's not going to work on a broader patient population.

"Again we were very good at going into broad patient populations. What we saw actually was that as the programme moved through the clinic, the commercial organisation went full steam ahead and wanted to go broader and bigger. Of course AstraZeneca was very much a primary cadre of an organization and so what we saw actually in the data was that the scientists were becoming less confident about their projects and the commercial folks were becoming more confident. But 100% of nothing is not a very big number." — Mene Pangalos

Right Commercial

Why would anyone want to take or prescribe the medicine and why would anyone want to reimburse it?

"Understand what your comparators need to be, understand what the standard of care will be in the time frame that you are going to be launching.

It's a very difficult thing to do, often 10-15 years ahead, but really challenge the teams to think about where that puck will be when the programme moves through the clinic or when it launches to make sure they are being ruthless about the comparisons they do. This now goes back to being outward-looking versus inward-looking. We used to be too inward-looking as an organisation."
— Mene Pangalos

3. REWARD YOUR SCIENTISTS FOR KILLING HYPOTHESES.

"Are your scientists asking killer questions to try, not just to validate, but to *invalidate* your hypothesis? We celebrate good kills every day. We are passionate about it!

Failing early is important because it means you haven't spent too much money and you don't just keep on. Before, we were very good at finding ways of getting to the next hurdle just for the sake of getting the next hurdle, because that's what we were being measured on..."

4. FROM PERSONAL BEST TO WORLD RECORDS.

"If you think of what we do as a competitive sport... our view of innovation was more of personal best rather than world records. But innovation should mean you are cutting edge, you make discoveries rather than follow discoveries.

We were internally referenced versus externally referenced. We were getting better internally but when your benchmark is very low, you're getting better on a very low benchmark; it isn't getting you anywhere near where you need to be.

One of the big shifts in our culture was being much more outwardly focused, seeing what's happening outside, where we should be pushing ourselves to be even better, who we should be working with, who's going to help us achieve what we want to achieve."

5. DOUBLE DOWN ON YOUR STRENGTHS.

"When Pascal Soriot joined the company as the CEO at the end of 2012 we focused down on the areas where we thought we could be globally competitive and set world records: oncology, cardiovascular, metabolic and respiratory diseases.

As we went deeper and more focused in those areas, we started to build a depth of knowledge and pipeline that made us competitive. The quality of our partnerships, the quality of the people we recruited, the decision making; it all got better".

6. MIND THE INCENTIVES.

"We are quite careful about incentives because it can drive the wrong behaviour. It's easy for people to start gaming whatever they're given as a target, and scientists are brilliant at doing that!

The rewards for R&D come from good quality work, good kills, inventive things, innovative things, demonstrated proof of mechanism, demonstrated proof of concept, diagnostic strategies, launches, and phase III investment decisions.

This means I don't get to decide what goes into Phase 3... Someone else has to put that through and so this way you can't game the system.

We also do full three-year holding averages, so no one is ever pressured into doing something in one year and getting a number."

7. WHEN DISCOVERY AND SCIENCE BECOME ONE.

"I had an organisation of about 5000 people when I joined and we were publishing about 200 papers and one Nature or Science paper.

Today we are about 2500 people, and we're publishing between 40-50 Nature Science papers a year. Now it's part of our DNA to do both good discovery and science. It's all the same thing for us.

As a consequence, people want to come and work with you, whether it's an academic collaborator, a biotech or someone who actually wants to be a part of AstraZeneca."

8. DON'T KEEP YOUR IDEAS TOP-SECRET.

"We were incredibly closed. We didn't want to share anything. Everything was proprietary. And to change that culture we had to do it in baby chunks... you chip away, you chip away, until eventually, people get comfortable.

I get irritated by people who hoard data or think that they can't share things. If I ever have to choose, I always prefer to be too open rather than less open. The risks are relatively small and the upside is huge."

9. KEEP YOUR DOORS OPEN.

"We had this huge site that was half empty and I used to wander through the corridors going from one group to the other and there were those empty labs (we called them tumbleweed labs) where you could hear the winds rushing through; it was demoralising.

So one day we said, let's collapse our footprint on the building and let's bring biotechs in, and in contrast to other bioparks, let's not have the biotechs partioned and walled off; let's have them using our cafeteria, coffee shops, shared spaces, and equipment if they want to.

And really try and share our infrastructure, make ourselves good partners, help give them some regulatory advice, some clinical advice, when they need it, without asking for anything in return.

This encouraged biotechs to come in, it helped us to forge relationships with other companies, but most importantly, it filled the space up and made us feel vibrant, energetic and full.

We had a half-empty building in Boston that's now packed and has a waiting list for biotechs.

The idea here is to treat people like grown-ups. When we first set up this culture people were like, 'what do you mean they're going to be wandering around?'

I mean... everyone signs a CDA, if they don't follow what they should be doing they'll get kicked off the side.

If we go in with the assumption that everybody is going to behave themselves and actually follow the appropriate principles, then you're pretty safe. You don't have to have barriers and passes and everything else."

10. MOVE WHERE THE BRIGHT MINDS ARE.

"Moving to Cambridge was part of our cultural shift. Our new building is next to the Addenbrooke's hospital, the Papworth hospital, and the MRC Laboratory for Microbiology that has more Nobel laureates than any other institution in the world.

In Cambridge, you go to a coffee shop and you bump into someone who happens to be a haematologist, and you start to talk about things that you couldn't talk about when we were in Cheshire. It's amazing how many collaborations have been initiated through these informal connections.

One of the things that I've been trying to do over the years is to try and generate as many opportunities for our scientists to have informal connections. You're just making it easier for the serendipitous to happen, and then innovation can happen..."

ps: As I was wrapping up this post, Frank David at Pharmagellan shared a Nature article saying that AstraZeneca updated its 5R framework to include Right Digital Tools in clinical trials (patient-centric endpoints, digital biomarkers, remote patient monitoring, dose selection, event adjudication with AI, study design review with AI, digital recruitment, digital integrated clinical trial solutions, remote visits and telehealth, devices and sensors, etc.)

If I were to add an extra R, I'd go with this: "When we submitted the 5R paper for review, one of the reviewers said, if you do all this you need to add a 6th R, the Right Culture. Because this is all about transforming the culture of the company. She was right.." — Mene Pangalos

11 STRATEGIES FOR MOTIVATING AND HOLDING A BIOTECH TEAM TOGETHER FOR A LONG TIME

"Drug discovery is an insanely complicated activity;
what makes a great leader in our industry is the ability to hold a
team together for a very long time."
—*Joshua Boger, Founder of Vertex*

1. DON'T TELL A STAR WHAT TO DO.

A biotech team is a bunch of brilliant PhDs, postdocs, and scientists. These folks are hustlers by nature, but you have to press the right buttons.

Rule #1: Smart people don't respond well to being told what to do. What you do is, hopefully, you inspire them to want to take action. And this is the difference between leaders and managers: Managers tell people what to do whereas leaders inspire them to do it.

Daniel Pink in his book Drive mentions autonomy as one of the three things that motivates people among purpose and mastery: "People need autonomy over task (what they do), time (when they do it), team (who they do it with), and technique (how they do it)."

2. HAVE EVERYONE IN THE TEAM TALK AND STAY CLOSE TO PATIENTS.

Take the time to leave the lab, the clinic, the office and meet with the patients of the disease you're trying to treat. It will give you so much energy and focus.

3. MAKE SURE THE VISION IS BOLD ENOUGH.

Stars won't work for any company; they want to work for a company that sets out to achieve the impossible, i.e. a new modality, a new class of medicines. Attracting and retaining talent always starts with a bold vision.

4. HAVE THE TEAM REFLECT ON THE VISION FREQUENTLY.

The daily grind can consume you. You need a spiritual practice as a team. Spend 5min every week after an all hands meeting reflecting on your why. Have different people talk about what the vision means to them. This spiritual practice will save you when you find yourselves at crossroads.

Teams often get divided in the face of critical decisions because the vision is not iterated enough, e.g. decisions like the one George Yancopoulos, the Founder of Regeneron, is talking about below:

"What most companies do is... at a very early stage when they really think they're onto something, they go to Pfizer or Amgen or Merck and they sell out, and the company is gone!

They get what seems to be a lot of money upfront, let's say $100m, the couple of guys who started the company make a decent amount of money... but the company is done! It's then all up to the big company, but what happens is ultimately most of these projects die in the big company...

5. MONEY DOESN'T MOTIVATE.

Back in 1995, Microsoft paid writers big bucks to write Encarta, an encyclopedia it sold on CD and as software. Ten years later, they had to close it down defeated by a competitor that paid no one and offered its encyclopedia for free: Wikipedia!

External incentives can't inspire people to care. As Wharton Professor Russell Ackoff said, "Money to a company is like oxygen to a human being. If you don't have enough, you have a problem. But if you think life is about breathing, you're missing the point".

The other problem with incentives is that they're easy to match, so they don't give your company a competitive edge for recruiting, retaining, and engaging top talent.

Be generous and reward your team for outstanding performance. But traditional "if-then" carrot type of rewards kill your team's intrinsic motivation. Don't waste those infinite reservoirs of energy waiting to be deployed to your bold vision.

6. KNOW EACH MEMBER OF THE TEAM DEEPLY.

How can you motivate them if you don't know what motivates them? Talk to them. 1-on-1. Soul-to-soul.

Tell them, "What is your personal vision? What are your aspirations and dreams? How can we help you achieve those dreams?"

Your job is to align what your company's trying to accomplish with what they're trying to accomplish. If you can match their unique skillset, intelligence, passion, and aspirations with your vision, you have succeeded.

And remember that people just want to talk to you to tell you how good they feel or how challenging times seem. If you have enough time for your people and you show you care about them, they pay it back 10x.

7. CREATE BUY-IN.

It is important for your team to be invested in a decision, otherwise, their execution will be half-hearted or won't even happen. You create buy-in when you make people feel that they're part of the decision. Not all perspectives can win but all perspectives must be considered to create buy-in. People want to feel heard. This applies to decisions, company values and vision.

8. TREAT EVERY EMPLOYEE AS AN A PLAYER.

Steve Jobs used to say about his mentor and coach Bill Campbell, "Bill can get A performances out of B players". But to achieve that, you first have to believe that a "B player" is capable of achieving A performances.

You have to believe that the person you thought of as a B player may actually be an A player. What would happen to their performance if you truly believed they're an A player? How would you treat them differently? How would they treat you and the task differently?

9. SHOW APPRECIATION. CONSCIOUSLY, DELIBERATELY, AND OFTEN.

Tell people how amazing they are; knowing is not enough. Add a weekly "love the team" block in your calendar. Write or talk to them one by one, not a "hey thank you for your hard work everyone" in Slack!

When you catch yourself feeling grateful about someone or something they've done, let them know. When you hear something nice said about someone, let them know. And be specific: "Jen, I appreciate you for updating those process documents".

10. ALWAYS LEAVE THE DOOR UNLOCKED.

Don't try to retain your stars by locking the door. Help them learn and grow, rotate them, invest generously in their training without any expectation. If they leave you, they'll tell everyone in the tiny biotech world how great you are.

They may have to go somewhere else to get the necessary skills and come back when the time is right to contribute 10x towards your vision. And remember... just like your vision and your culture, your alumni will help you win the long war for talent.

And last, but most important…

11. DON'T TRY TO BE A RAMBO TYPE OF LEADER.

You don't have to always show up confident and powerful in front of your team. You shouldn't. You must not!

Motivation is like a feeling; it comes and goes. This is where the leaders who want to do a great job fail: expectations.

There is no way you and your team can feel motivated 24/7. Feeling disappointed because you didn't get the expected data? Worries about fundraising? Great!

Don't spend infinite amounts of energy to hide those emotions from your team and show up strong. Because you make them more anxious when they can see a gap between what you feel inside and what you project outside. The dissonance is what makes people feel unsafe.

Slow down to acknowledge those feelings. You must let those feelings out first before you can create space for motivation. And practice this as a team, "Disappointing data, I know. Feeling like s*** today like everyone else. BUT I'm confident we'll figure this out guys."

All your team needs to feel pumped again is that last bit: "We'll figure this out." Take a moment as a team to grieve the data you didn't get, accept, and then move onto action.

20 TIPS ON BECOMING A GREAT MANAGER

If biotech is a team sport, the CEO or manager must be a master coach. If I had to recommend just one book to aspiring master coaches, it would be High Output Management by Andrew Grove, the former CEO of Intel.

20 key ideas from one of the greatest CEOs and management teachers:

As a CEO or manager you have to be *obsessed* with the output of your people.

Your Output = Their Output.

You can dive in, do stuff, and be heroically productive but that's not going to maximise your output. Your job is to help everyone else do a great job.

A manager's output = The output of his/her organisation + The output of the neighbouring organisations under his/her influence.

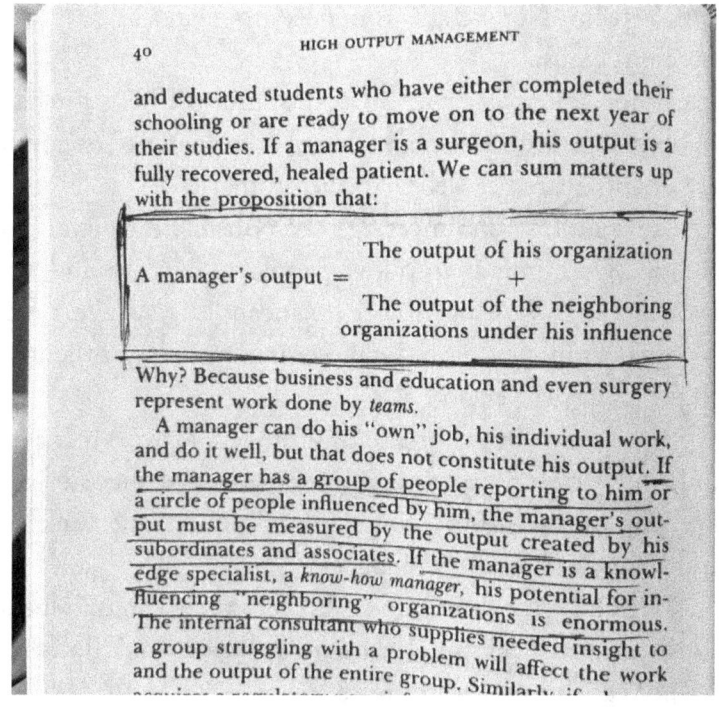

1. Managerial leverage is the impact of what managers do to increase the output of their teams. High managerial productivity depends on choosing to perform tasks that possess high leverage.

2. A manager's work involves allocating resources: manpower, money, and capital. But the single most important resource that we allocate from one day to the next is our own time. How you handle your own time is the single most important aspect of being a role model and leader.

3. Your time is your one finite resource, and when you say "yes" to one thing you are inevitably saying "no" to another.

4. No amount of formal planning can anticipate changes within a business's environment. Does that mean that you

shouldn't plan? Not at all. You need to plan the way a fire department plans...

It cannot anticipate where the next fire will be, so it has to shape an energetic and efficient team that is capable of responding to the unanticipated as well as to the ordinary.

5. The chairman of a meeting is responsible for maintaining discipline. It's criminal to allow people to be late and waste everyone's time. Wasting time here means that you're wasting the company's money, with the meter ticking away at the rate of $100 per hour per person.

Do not worry about confronting the late arriver. Just as you would not permit a fellow employee to steal a piece of office equipment worth $2,000, you shouldn't let anyone walk away with the time of his fellow managers.

6. When a person is not doing his job, there can only be two reasons for it. The person either can't do it or won't do it; he is either not capable or not motivated.

To determine which, we can employ a simple mental test: if the person's life depended on doing the work, could he do it? If the answer is yes, that person is not motivated; if the answer is no, he is not capable.

7. The old saying has it that when we promote our best salesman and make her a manager, we ruin a good salesman and get a bad manager. But if we think about it, we see we have no choice but to promote the good salesman.

Should our worst salesman get the job? When we promote our best, we are saying to our subordinates that performance is what counts.

8. Given a choice, should you delegate activities that are familiar to you or those that aren't? Before answering, consider

the following principle: delegation without follow-through is abdication. You can never wash your hands of a task.

Even after you delegate it, you are still responsible for its accomplishment, and monitoring the delegated task is the only practical way for you to ensure a result. Monitoring is not meddling, but means checking to make sure an activity is proceeding in line with expectations.

Because it is easier to monitor something with which you are familiar, if you have a choice you should delegate those activities you know best.

9. Eliciting peak performance means going up against something or somebody. For years the performance of the Intel facilities maintenance group, which is responsible for keeping our buildings clean, was mediocre, and no amount of pressure or inducement seemed to do any good...

We then initiated a program in which each building's upkeep was periodically scored by a resident senior manager, dubbed a "building czar." The score was then compared with those given the other buildings. The condition of all of them dramatically improved almost immediately...

Nothing else was done; people did not get more money or other rewards. What they did get was a racetrack, an arena of competition. If your work is facilities maintenance, having your building receive the top score is a powerful source of motivation.

This is key to the manager's approach and involvement: he has to see the work as it is seen by the people who do that work every day and then create indicators so that his subordinates can watch their "racetrack" take shape."

10. Once someone's source of motivation is self-actualization, his drive to perform has no limit. Unlike other sources of

motivation, which extinguish themselves after the needs are fulfilled, self-actualization continues to motivate people to ever higher levels of performance.

11. Managerial meddling stems from a supervisor exploiting too much superior work knowledge (real or imagined).

The negative leverage comes from the fact that after being exposed to many such instances, the subordinate will begin to take a much more restricted view of what is expected of him, showing less initiative in solving his own problems and referring them instead to his supervisor.

12. You've learned that a valued employee has decided to quit. In such a case, you have to act fast to change their mind. If you put it off, your chances are lost. Timeliness here has very high leverage.

13. Don't put off decisions that will affect the work of the team. The lack of a decision is the same as a negative decision. No green light is a red light.

14. If you still want to keep and not delegate certain activities that you enjoy doing, that's okay. You deserve to and should do stuff that gives you energy but do it consciously.

15. When the environment changes faster than one can change rules, or when a set of circumstances is so ambiguous that a contract between the parties that attempted to cover all possibilities would be complicated, we need another mode of control which is based on cultural values.

16. In the end self-confidence mostly comes from a gut-level realization that nobody has ever died from making a wrong business decision, or taking inappropriate action, or being overruled. And everyone in your operation should be made to understand this.

17. My day always ends when I'm tired and ready to go home, not when I'm done. I am never done. Like a housewife's, a manager's work is never done. There is always more to be done, more that should be done, always more than can be done.

18. Is it better to be a hands-on or hands-off manager? It depends. If the employee is immature in the task, then hands-on training is essential. If the employee is more mature, then a delegate approach is warranted.

19. In order to build anything great, you have to be an optimist, because by definition you're trying to do something that most people would consider impossible.

#20 is not another quote but a deep realisation of mine from reading this book multiple times and living it with my clients. There's no single mention of the word "love" in this book but Andy Grove wrote this book with so much love—the same love that he had for his people.

We often look for magical tools and techniques to motivate our people but no intellectual trick can replace a leader's love for his people, for the vision, for the impact—and in a biotech context—for the patients. When you truly care, it's easier to mobilise people.

WHAT A BIOTECH LEADER CAN LEARN FROM THE CULTURE BAXTER INTERNATIONAL

299 biotech companies went public between 1979 and 1996 in the US. Of the top executives of those companies, 81 came from Baxter International. Henri Termeer was one of them. How did Baxter's culture breed exceptional biotech leaders?

1. SINK OR SWIM.

Imagine that you're a bright 28-year-old MBA graduate working for a top medical supply firm. Suddenly, on a Friday morning, you're told you'll be running a division in another country on Monday—with a flight booked for this evening. Sounds outrageous, right? But at Baxter, this was the norm. Young and inexperienced employees were thrust into positions of immense responsibility without warning.

William Graham, Baxter's CEO from 1953 to 1980, believed in throwing his employees into deep water and seeing if they sank or swam. If anyone hesitated, Graham saw it as a lack of self-confidence, affecting their career path. Baxter expected its employees to solve problems, handle people, and manage crises.

By pushing them out of their comfort zones, the company instilled remarkable confidence in its managers. They discovered their ability to learn on the fly and thrive, no matter the situation.

"We developed tests for Chagas disease based on feedback indicating that it would be a big market," Henri Termeer recalled. "Baxter asked me to head up this project. They said, figure out a way to set up the connections. That was a very Baxter thing to do..."

2. LEARN ON ALL FRONTS.

Baxter exposed their people to diverse situations, offering numerous educational opportunities.

While other companies had narrow specialists, Baxter groomed managers with a broad knowledge base, making them versatile and well-rounded.

3. SEEK MENTORS AND UNCOVER YOUR BLIND SPOTS.

Managers paired neophytes with seasoned veterans to develop nurturing, mentoring relationships. New employees were encouraged to ask questions, recognise their own limitations, and seek the advice of more experienced individuals.

4. ASCEND TOGETHER.

Baxter managers had to work hard and climb the ladder, or risk being let go. Despite Baxter's highly competitive atmosphere, its managers shared an atmosphere of camaraderie and friendship.

In the 1980s, ex-Baxter managers gathered for "alumni reunions," with over 100 attendees.In 1999, numerous Baxter alums joined a formal event celebrating Graham's 80th birthday. Graham spoke proudly of those whose careers he had influenced so profoundly.

5. ALWAYS CREATE AND MAINTAIN RELATIONSHIPS.

A key aspect of Baxter's philosophy was the cultivation of extraorganisational and government-based relationships. Also, unlike companies where people who leave are regarded with resentment, the top brass at Baxter maintained connections with its former employees. With only a limited number of senior management positions, Baxter's leaders understood that turnover was inevitable.

They also realized that cultivating relationships with influential former employees would enhance Baxter's eventual entry into the biotech industry.

Now... you don't have to be like Graham or have a Baxter culture to be a great leader who fosters both personal growth and business success. What matters is crafting your unique leadership style and values. Because if you're not you, even you won't follow you. Having said that, qualities such as commitment, hard work, adaptability, resilience, relationship building, and resourcefulness.. are absolutely necessary, particularly in a biotech context; you have to instil them in your culture.

JON SKVARKA'S LEADERSHIP PRINCIPLES

Jan Skvarka took Trillium Therapeutics from a $16 million market capitalisation to a $2.26 billion acquisition by Pfizer in only 2 years and in the midst of the pandemic.

When Skvarka took the reigns at Trillium in September 2019, the stock price had fallen from $30 to $0.40, the company was about to be delisted from NASDAQ, and they had 5-6 months left of cash.

Given the challenging situation, Skvarka spent six months on the diligence of the underlying science and Trillium's assets. He wanted to confirm that the scientific foundation was solid and that he had the board's support.

Before he joined, he wrote a memo to the board outlining his analysis of the situation, strategic options, and a plan and hypothesis for how to proceed.

Once he had the board's support, he was in.

Skarkva described Trillium's 2-year turnaround in three stages:

1 - The Emergency Room - 3 months

"We approached the situation like a patient in an ER; we first had to stop the bleeding. We laid off 40% of the staff and reset our strategy focusing on our most promising assets." Decisiveness and speed helped us rebuild investor confidence.

2 - The Transformation - 15 months

"It was a lot of systematic blocking and tackling, touching every aspect of the company. We had a new team in place, a restructured executive team, a restructured board, a world-class SAB, and two promising molecules.

We did a lot of work in the trenches with investors. We had over 400 meetings with investors in 2020. We raised $300 million in two rounds and secured an equity deal with Pfizer for $25 million."

3 - The Build-Out

"With money in the bank, we launched a broad Phase 2 program with nine clinical studies across multiple indications in both hematologic malignancies and solid tumours, while doubling the size of the organization within the next year.

We were in a great place to continue building the company, and we had high confidence in our assets. However, come May 2021, the stock price started coming down, due to the markets getting nervous about biotech and some retail investors selling out of their Trillium positions.

I approached Pfizer's head of business development for oncology, to ask if Pfizer would consider another $25 million equity investment, given that we had generated substantial additional data since Pfizer's initial investment.

Pfizer asked to look at the data and started the diligence. While they were looking at the updated Phase 1 results, Phase 2 data started coming in, including some from very tough multiple myeloma and AML patients, two indications of strategic interest to Pfizer.

It was at this point that Pfizer shifted their interest from a simple equity investment to a full buy-out..." From an initial bid of $15 per share, Pfizer ended up buying Trillium at $18.50...

6 Lessons:

1. DON'T PARTNER JUST FOR THE MONEY.

When Skvarka first partnered with Pfizer, he added the CSO of Pfizer Oncology to his SAB.

From that point, Pfizer had a front-row seat to see what Skvarka was doing, they liked what they saw, and they ended up pulling the trigger.

"The partnership with Pfizer was not just about the money; it was about validating our work, building relationships, and getting access to Jeff's expertise". — Jan Skvarka

2. LAYOFFS ARE PAINFUL BUT "YOU HAVE TO STOP THE BLEEDING."

"The toughest part of my time at Trillium was the Emergency Room. Turnaround situations require decisive leadership and speed. There is no room for hesitation. You cannot procrastinate.

Maybe you can still pay severances if you move today, but if you prolong things six months the company could run out of cash. And the team is deflated and maybe considering moving on to other opportunities — so staff are already anxious to hear where we are heading.

So, we did those changes with speed, balanced with respect and transparency. It was this respect and transparency with our staff that allowed us to rehire several laid-off employees once we were stabilized."

In a video about Trillium's success, a former employee said: "Jan had to take some difficult decisions, I was in that 40% that he had to let go" (Not sure if that person was rehired, but regardless, this speaks volumes about the culture and letting people go with respect).

3. CULTURE IS NOT ABOUT BUZZWORDS; IT'S ABOUT ACTIONS.

"I'm not a believer in lengthy meetings about culture and putting lots of buzzwords on the wall. Buzzwords are cheap — after all, even Enron had "Integrity" engraved on their headquarters entrance wall.

I think culture is shaped organically through how we make decisions, who we recruit, and how we treat employees. It has to be with respect, transparency and honesty, while always keeping patient impact in mind."

4. HAVE A CLEAR DIRECTION.

"It's critical to have a vision and be crystal clear about the company's direction. And not just clarity about who you are, but who you are not. Without this clarity, companies and executives get distracted and pulled in different directions."

5. PHARMA PARTNERSHIPS ARE ALL ABOUT TRUST.

"Trust is key. Don't oversell, always be honest and transparent. In our case, Pfizer had about 150 people on the due diligence, and they turned over every stone.

So, if there are any problems, just be transparent about them, and disclose them upfront, as you don't want them to come up later in the diligence and then play defence."

6. ALWAYS REMAIN HUMBLE.

Skvarka said, "Were we successful? I am not declaring success. The success will come when our CD47 molecules are on the market and they're saving patients' lives".

BIOTECH FOUNDERS TALKING ABOUT HUMILITY

Humility is probably the most critical quality for success in your arduous journey of bringing breakthrough medicines to patients.

I am an avid reader and what has caught my attention is how often the concept of humility comes up in biotech founder and CEO interviews.

"The job itself will humble you. You'll be told no, you'll be told you're stupid, you'll be told all sorts of things. You have to be ridiculously humble and constantly be asking other people, checking on yourself, finding experienced folks, building a personal advisory board... I have a number of people... fellow CEOs who are at our stage, people who are a few years in the future, people who have done companies multiple times... that I can ask questions to." — Jake Becraft, Co-founder at Strand Therapeutics.

"You may find it daunting to ask for help, you may see it as if you're looking weak or you've failed in some way. The sooner you can get over being afraid of rejection or hearing no, the easier this will become. Because you'll get it from so many angles, i.e. fundraising, recruiting, sales..." — Lex Rovner, CEO at 64x Bio

"Being humble is at the core of developing managerial skills. While you don't have to follow all the advice you receive, it is important to stay genuinely open and sincere to all feedback." — Martin Trevor, Co-founder and CEO at Mammoth biosciences.

"We are taking the first approach to use mRNA therapeutics as programmable epigenetic medicines. It's a completely new class of drugs never before been attempted. And now we are entering the clinic... You have to have a significant amount of humility to appreciate that you're going to actually get patients to use these drugs. I mean, imagine the social contract that you need to have with patients, morally, to be able to make it." — Mahesh Karande, CEO at Omega Therapeutics

"Science is very humbling, and I've had far more failures than successes. I've gone into work many more days to have disappointing data presented to me, which sometimes meant the end of a program... If after that process, you're not humbled, I don't know what's wrong with you, frankly! There are people who really believe, okay, it's my idea and we're going to do it my way. That may work for a while but not in the long term." — Jim Sullivan, Co-founder and CEO at Vanqua Bio.

"No one is ever ready to be a founder CEO. There will always be more knowledge to acquire to be "ready". However, I believe there is a point at which a potential founder is ready from a mentality perspective. And the mental perspective is one of humility. I have found this shows up as folks (no matter the background going in) who recognize they have much to learn and are willing to do whatever is necessary to improve themselves." — David Li, Co-founder and CEO at Meliora Therapeutics

"You have to have the humility to listen to your team; they are often closer to the problem and have better insights than you." — Ramji Srinivasan, CEO at Teiko Bio.

"Humility is an under-appreciated skill in founders. My interdisciplinary training gives me an edge in understanding how computation and lab experiments can reinforce each other when tightly integrated. In all other areas, there are people in the company who are smarter and better understand the details of our processes. Never let ego get in the way of the right decision." — Nicolas Tilmans, CEO at Anagenex.

"Drug development is a hard task and daily leadership is about humility in front of the task you have. We tackle the hardest task and you have to be humbled in front of that." — Stan Crooke, Founder and Chairman at Ionis.

"Biotech success is very hard. 99% fail if you're in preclinical stages. I use the baseball analogy... The best hitters of all time, still fail more than 50% of the time. If you have a battling average of 0.333, 0.350, that's Hall of Fame. The same is true with drug development. That should keep people very humble. Every time you take on a new product, a different drug, a new target, a unique modality, a specific dose going after a specific indication... everybody is always doing this for the first time! I don't care if they have 50 years of experience, every time you take on a new company, a new programming product, you're doing it for the very first time. So you need to keep that humility. It's really about going through all the permutations in your mind, all the scenarios, not waiting for the problems to happen and then trying to fix them, but anticipating all the different things that can go wrong. What don't we know? What can we learn more?

How do we get a better understanding of what the right dose is? Can we do some additional experiments? It's striving to derisk more and more before you push something into patients. That humility solves for a lot of all the other things required from the CEO, i.e. being team-oriented, having a great culture, etc. so that's why I believe it's the most important quality. — Chris Garabedian, CEO at Xontogeny.

THREE THINGS THAT INVESTORS WANT TO SEE FROM YOU—A SUCCINT SUMMARY

An incredibly succinct summary by one of my good mentors Janis Naeve, a Partner at Cota Capital, who has 20+ years of experience in biotech investing.

1. THE TEAM.

It's all about the team, i.e. the founders, the experience you have in the space, your unique and complementary skills, and the teams you've put together in the past...

"I don't want a team who are going to tinker around... I want a bold, committed, and undaunted team... On my first investment, I lost money because I fell in love with the science and didn't consider the team".

2. THE PROBLEM.

If you guys are able to accomplish what you set out to do, how important and significant is that? Is the juice worth the squeeze?

3. THE PLAN.

Okay, you're a strong team working on an important problem. Can you come up with an operational plan and budget that make sense? Can you translate your big vision into points along the way and be able to measure your progress or pivot if necessary?

To me guys, the point here is not just about convincing your investors. Being able to navigate fundraising shows a lot about you and the business, but the Odyssey starts after fundraising, i.e. when you have to hit all those milestones and demonstrate operational excellence.

The three points above are about setting high standards for your company and increasing your chances of success in the challenging journey of bringing life-saving medicines to patients.

12 TOOLS FOR BUILDING A HIGH-PERFORMING BIOTECH TEAM

All my biotech CEOs/execs have struggled with this one thing: Team Efficiency and Speed.

How do I make my team go faster? How do I create urgency? Iteration speed is slow.

1. EXTREME OWNERSHIP

You are the CEO. It's entirely your responsibility to increase the output of your team. It's #1 item in your job description. You've hired a bunch of A players; now you have to make them a team. Your CEO output = Their collective output.

2. IT'S NOT A PROBLEM; IT'S BAU.

It's not that you woke up one day, you noticed some unexpected inefficiency issues within your team which you had to fix, and moved on.

Building a high-performing team is a forever work in progress. No team was a dream team from Day #1. No leader was a dream leader from Day #1.

3. BE VULNERABLE, OPEN, AND CURIOUS.

Say to your team, Guys, I sometimes feel like we could be going faster but only you can help me see if this is true. [Define what "faster" or "more efficient" means here].

It's either my unrealistic expectations (which will always be true to a good extent :) or I'm not doing a good enough job to support you. What do you say?"

How can we, if we can, become more efficient as a team?

What could be slowing us down?

Do you have all the help and support you need?

How can I, or someone else in the team, support you more?

4. ASK OPEN-ENDED QUESTIONS.

Notice that all the above questions are open-ended coaching questions, i.e. they cannot be answered with a yes or no.

Don't assume you know the reasons that are slowing your team down. You may have some ideas but you don't have the full picture. They are the ones closer to the work, so invite them to wonder, be curious, and "meditate" on efficiency by asking the right questions.

5. TALK TO EVERYONE ON THE TEAM ONE-ON-ONE.

Before you talk to the whole team, spend a good time talking to people one-on-one. Not everyone will be as comfortable sharing their observations or concerns in an all-hands.

5a. Ask about their individual performance without being judgmental.

Focus on facts (what you can see) and avoid evaluations. You don't have the full picture anyway.

5b. Solicit honest feedback.

Tell them that you're working on becoming a better leader with the help of your coaches and mentors and that you're learning alongside everyone else.

If they need more support from you, ask, "What am I not doing to help you do your job more efficiently? Help me become a better manager!"

5c. Ask them about the whole team.

How can we become a more efficient team? (See questions in #3 above)

5d. When they seem to have given you all the reasons, don't move straight to solutions.

Ask the magic question: "What else?" "Is there anything else that you haven't mentioned?" You have to dig deeper to find the real issues.

Remember that the most important job of the CEO/ manager is information gathering. You're a coach for them. Your commitment is to understand your people deeply. These 1-on-1s is the most powerful management tool and the best investment of your time.

6. FEAR OF FAILURE: COACH THEM OUT OF IT!

It always comes down to fear of failure in my experience...

You can't expect your team to generate breakthroughs or iterate faster if you haven't given them permission to "fail".

A CEO said to me, "My team wants to celebrate the things that work!" Of course, they do! Everyone does!

But how can a biotech team truly innovate without being unreasonable and comfortable with"failing"?

This means that you, the leader, have to do colossal work to redefine failure in your culture. It's conscious, deliberate and consistent work.

And you have to do even more work with team members who have little work experience or joined the company straight out of academia.

Imagine it's your first gig in the industry. It's a new world! Everyone around you is smart, talented and an expert in their field.

You want to prove yourself, you want to generate results, you want to impress—which is fine!

The problem is that this can force people to play safe, i.e. go after the "easy" hypothesis/experiment and chase what's most likely to *work*.

The role of the leader here is critical! You can't just say, "Hey guys we are true innovators here, I'm okay with you failing", put the word "unreasonable" on the wall, and expect them to go and break the lab the next day.

You have to do a lot of work to breed innovators.

6a. Focus on learning, not just winning.

It doesn't matter if we don't get the desired data. What can we learn from this experiment? Why is it crucial to cross this off? What can we do next if this doesn't work?

6b. Create a metric for unreasonable bets.

In all-hand meetings ask everyone to talk about an "unreasonable thing" they tried, no matter whether it worked or not. What did you learn? Talk about your own unreasonable bets and lead by example.

6c. Pay attention to their words.

If they only say, "I should do x. I know what I have to do", then they may be not taking enough risks. Nudge them, "Ok... other than this, is there something else that you'd like to try, no matter how crazy or impossible it may sound?"

6d. Be mindful of what you praise.

If you *unconsciously* praise only what works, they'll make sure to only do things that work! If you only want to hear "good news", they'll make sure you get good news!

6e. Ask the team to read Mindset by Carol Dweck.

Once you've all read it, discuss it and apply the principles.

This book will help you and the team move from a know-it-all (fixed mindset) to a learn-it-all (growth mindset).

7. INTRODUCE A GOOGLE 20% TYPE OF RULE.

The famous Google 20% rule was introduced in 2004 and allowed employees to dedicate 20% of their work time to pursue personal projects, fostering creativity and innovation within the company. This is how Gmail was born.

Whether the Google 20% rule still holds today is a debate, but it doesn't matter. The point is, introducing a rule that allows people to spend some time "going wild" sends a clear message about the culture in your company.

8. CREATE A "SHOCKING RULE".

In the early days of Facebook, when Mark Zuckerberg was trying to catch up with MySpace, speed was the #1 virtue he needed from his team, so he created a shocking rule: "Move fast and break things".

Imagine you're a software engineer obsessed with your "clean code" and you hear your CEO say: Break things! Well, what the rule says is... "Have you got something cool in your mind? Push it along and don't think too much! Right now, we need speed."

So the rule has to be a little bizarre/stretched making people go..."What???"... which means they won't forget it. Obviously, you can't break things in biotech but can you come up with a rule/message that can align and rally the team right now?

Here's an example of a rule around strategy that the former CEO of Alnylam, John Maraganore, used to rally his team in a critical time for the company.

"We discussed communicating a set of five-year goals at the upcoming January 2011 J.P. Morgan conference. I wanted to

propose a new strategy called 'Alnylam 5×15', with a commitment to advance five RNAi therapeutic programs into clinical development by the end of 2015.

These programs would all be focused on liver-expressed, genetically validated disease targets (where we had achieved reliable delivery results in primates). In addition, we would focus on targets for which human POC could be realized as early as phase 1, based on biomarkers, and we'd create pivotal studies with endpoints meant to support regulatory approval and demonstrate value for payers. The team pushed back on this. With only one liver-targeting program in development at that time (ALN-TTR01), and no human POC data in hand, the team suggested that '2×15' or '3×15' might be a more manageable goal to promote publicly. I disagreed, saying that a reduced number of targets would not excite our stakeholders nor motivate our team. And, so, Alnylam 5×15 it was! Our research organization rallied behind the new strategy."

9. APPRECIATE THEIR PACE AND COACH THEM INTO GOING FASTER.

You may know from experience that a certain task shouldn't take someone more than a week to complete. You have also outlined the subtasks. Still, don't say, "I think you can finish this in a week".

Ask them, "How much time do you need to complete this?" Give them as much time as they need. When they're done, you can ask, "Do you have any ideas about how you can do this even faster next time? What did you observe?"

I asked a CEO, "Imagine you are them for a moment. How are you feeling when the CEO says, you can finish this in a week".

"I would feel pressured", he replied. Well, rightfully so...

10. DELIGHT YOUR CUSTOMER.

In biotech, you get this kind of interdependencies where, on one hand, the dry lab can't do its work until the wet lab is done... And the wet lab can't improve the next experiment until the dry lab has analysed the data of the previous experiment.

An interdisciplinary team is powerful but when you have this kind of interdependencies, speed can suffer. And speed is critical when you're not big pharma but a small company with a limited runway.

Say to the team, the person you hand off your work to is your customer. How can you delight your customer? What is important to them? What are their requirements? Often best practices in one team may create problems in another team further down the line.

A team may have to compromise its standards for the next team or the common good. Remember the classic from Systems Theory: "to optimise the system, you have to sub-optimise the sub-systems."

11. INVITE THEM TO STEP IN YOUR SHOES.

Tell them what's going on in your world as the CEO/leader. Your eyes may be fixated on the runway. You're getting pressured by the board and investors. Your people will do anything to support you when you're open and honest with them.

12. GIVE THEM MORE CONTEXT ABOUT THE BUSINESS.

A company is a living organism. The corporate strategy and the business environment are constantly changing. It's not enough to tell people "we have to go faster now".

You have to give them context. What's different now? Why are you asking me to stop doing x and start doing y all of a sudden? If you help them understand the situation, not only will they be more motivated to run faster but they'll provide you with more insight and solutions.

BUILDING A HIGH-PERFORMING BIOTECH TEAM - SUMMARY

1. **Own it.** It's the CEO's/manager's job to increase the output of her team.
2. **It's not a problem.** Increasing output is a forever work in progress. It's your job.
3. **Be vulnerable and open.** Guys, what can I do better for us to become a better team?
4. **Ask open-ended questions.** You don't know what's slowing them down. Remain curious.
5. **Talk to everyone one-on-one first.** (Individual performance, solicit feedback, team performance, dig deep to find the real issues)
6. **Coach them out of fear of failure.** (Focus on learning, encourage unreasonable bets, read the fear behind their words, don't only praise success but effort towards the right direction, read Mindset by Carol Dweck)
7. **Introduce a Google 20% type of rule.** Allow people to spend some time every week pursuing their own ideas or projects.
8. **Create a shocking cultural rule.** If speed is now the most important thing, can you create a memorable rule about it?

---

9. **Appreciate their pace and coach them into going faster.** Don't tell them how fast you expect them to run but coach them into running faster and improving.

10. **Delight your colleague-customer.** The person they hand off their work to is their customer. How can they serve their customer? What is important to them? What are their requirements?

11. **Invite them to step in your shoes.** Tell them what's going on in your world. Talk to them about your concerns and fears.

12. **Give them more context about the business.** Why is it important to run now? What's different today than yesterday? They need to know to buy in.

7 MUST-READ BOOKS FOR THE BENCH SCIENTIST TURNING BIOTECH LEADER/CEO.

1. FROM BREAKTHROUGH TO BLOCKBUSTER — THE BUSINESS OF BIOTECHNOLOGY BY DONALD AND LISA DRAKEMAN AND NEKTARIOS ORAIOPOULOS

This is a fantastic book/crash course in the business of biotech. You've got everything there, i.e. the players, the interactions, the agendas, the numbers, the challenges.

It's backed with data, it's beautifully written and it was only published in April 2022.

2. LETTING GO BY DAVID HAWKINS

Being a first-time CEO comes with a full range of uncomfortable emotions!

This book will teach you emotional resilience.

I've worked with hundreds of leaders; all end up becoming masters of the techniques explained in this book.

Letting Go will help you become more like a palm tree than an oak tree. The oak tries to resist the storm and ends up breaking, whereas the palm bends and comes back up again... The biotech journey is always full of storms and you have to be a palm tree.

3. NONVIOLENT COMMUNICATION BY MARSHALL B. ROSENBERG

This book gives you a framework on how to navigate challenging conversations with anyone, i.e. cofounders, investors, employees, or board—and do it with confidence. It will teach you how to speak your truth without the fear of "ruining the relationship".

4. HIGH OUTPUT MANAGEMENT BY ANDREW GROVE

Graduate training doesn't prepare you for managing people. This book is management explained in the most elegant and scientific way. This is where management meets leadership... or where science meets art! The former CEO of Intel, Andy Grove, the most loving teacher, left this gem to all of us.

5. TRILLION DOLLAR COACH BY ERIC SCHMIDT, JONATHAN ROSENBERG AND ALAN EAGLE

This is the leadership playbook of Bill Campbell, the coach who coached Steve Jobs, Larry Page, Sergey Brin, Eric Schmidt, Sheryl Sandberg, and Ben Horowitz, among others.

In a biotech company, you have a bunch of stars. How do you manage egos? How do you build an envelope of trust? How do you become a coach like Bill for your people? Although it was not written specifically about biotech, this is what this book is all about. There are so many great management tools in there. I give it to all my CEOs.

6. VENTURE DEALS BY BRAD FELD AND JASON MENDELSON

Biotech is a capital-intensive industry with years of loss-making operations... This is why Stelios Papadopoulos, aka the "Godfather of Biotech", has said: "Until you get a product on the market, all you've got to sell is stock!"

The biotech CEO and every member of the management team have to be incrementally better at fundraising and be able to tell a clear and compelling story to the financiers.

As the authors of the book say, "You have to be smarter than your lawyers and venture capitalists!"

7. CONSCIOUS BUSINESS BY FRED KOFMAN

This is my #1 book on company culture and team building.

You biotech folks are fighting against the odds and are called to tackle the most complex and difficult problems.

If some company out there can "make it" without passing the culture exam, you simply can't. You can't afford to play this long-term game without a great company culture. There's no room for friction with you guys. You all have to work together like a fist against the disease. Having a healthy company is an absolutely necessary condition for solving the biggest problems in human health.

10 PRINCIPLES FOR REAL MULTIDISCIPLINARY TEAMWORK

Having a multidisciplinary team and doing real multidisciplinary work are not the same thing.

Bragging about your ultra-diverse team means little if they never debate or challenge each other.

Here are 10 principles for doing real "mud scientist" work:

1. ESTABLISH A MATRIX ORGANISATIONAL STRUCTURE.

Yes, you should have reporting lines but in terms of how the actual work gets done, make sure that everyone has skin in the game, in everything that's going on in the company.

2. RECRUIT "MATRIX EXECUTIVES".

During interviews, beware of executives who're all about "how many people reported to me or will report to me" and "my team vs other teams". You can spot someone who wants to build a team around him or her. Pass.

3. CULTURE IS MODELLED AT THE TOP.

The leader sets the tone and the culture. If you don't like to be challenged, they'll pick it up and... they'll simply never challenge you. Soon the cultural rule will become: "To survive here, stay in your lane."

Tell the team: "You are experts in what you do. You know better than me. If you don't challenge me, you simply don't respect me."

4. WHEN YOU GIVE YOUR OPINION, ATTACH PHRASES THAT PROVIDE A PLATFORM FOR THE TEAM TO QUESTION YOU.

Say to your team, "Okay, now let's see if someone can poke holes in this. Tell me what's wrong with this idea."

5. BEST IDEA > CONSENSUS

Consensus means optimising for individual egos. Best Idea means optimising for the collective ego.

Tell the team: "We hate consensus here! Our goal is to always unearth the best idea. Anyone around is equally capable of spitting out the best idea. You guys either speak up or we go under!"

6. CREATE EXPLICIT CULTURAL RULES FOR SPEAKING UP AND SHARING DIFFERENT PERSPECTIVES:

"There are no stupid questions here. We challenge the status quo. We think from first principles. Remaining quiet undermines the team, the company and our mission."

7. SOLICIT INPUT ONE PERSON AT A TIME.

Don't just ask, "So, what do you guys think?" Asking everyone is the same as asking no one.

Go round and ask people one by one, "Keith, what do you think we should do here? Amir, how do you see things?"

8. PAY ATTENTION TO PEOPLE WHO REMAIN QUIET DURING MEETINGS.

The ones who often have the best idea are usually the ones who remain quiet because while everyone else is talking they're the ones who're thinking! Pass the ball to them, and again, one by one.

9. SIT DOWN WITH INDIVIDUALS BEFORE TEAM MEETINGS TO FIND OUT WHAT THEY'RE THINKING.

This will help you understand the different perspectives but most importantly give people come to the meeting prepared to talk about their point of view.

10. FIRST IMPRESSIONS LAST.

The onboarding process is the best opportunity to show your collaboration culture. Create a slide deck that gives a complete picture of the different teams and how everyone's work is connected. Plus, have new employees talk to all teams.

WHEN THE CEO TALKS TO YOUNGER SCIENTISTS

I asked a veteran biotech CEO what was one of the biggest lessons he had learned in his career.

He said, "I love talking to the more junior people in the company. I get a lot of pleasure out of that. I love coaching them and brainstorming with them.

But, when the CEO throws out opinions to a junior person, these opinions are highly impactful.

And there could be two layers of managers between you and that junior person. So, if you get too hotheaded with the scientist, then you're really disempowering their manager, or their manager's manager—and their managers get really pissed off with you!

You'll say, well, I just wanted them to know that the CEO cares. I wanted to show them that the CEO understands a little bit about what they're doing, etc.

Sure, you can do that but you have to be clear about the conversation.

You can say, I'd love to meet with you and grab a cup of coffee, but we are not making decisions here, we are only brainstorming. And you (talking to the scientist) have to go back to your boss and make that decision before you ever come back to me with a recommendation or anything.

As a CEO, you have to be very careful respecting all the management layers in between."

HOW TO ESTABLISH YOUR COMPANY VALUES

Establishing your company values is an extremely important exercise. Here are three considerations from my experience working with CEOs on this extremely important exercise:

1. Getting everyone involved. It certainly takes less effort, time and other resources to sit down as a senior leadership team to come up with the values of your company. However, involving all employees in this invaluable exercise can be extremely beneficial. The CEO can email everyone and ask them, we want to do this exercise, what are the values you want to identify with?

Asking everyone has the benefit that you create massive buy-in in the company. Everyone got asked, everyone contributed to forming the values, which means they will be willing to fight harder for those values.

You can reply, you can ask follow-up questions, and bring it up in 1-on-1's with people. Yes, it will take longer and more effort (sometimes it can take a company a year to complete this exercise!!!) but it's definitely worth it! It's a truly amazing opportunity to bring the whole team together, uplift the spirit, and boost motivation.

Once you collect everyone's input, you can refine it, and organize it into themes. You may start with fifty values and cut them down to ten or seven in the end.

2. These have to be *your values* that distinguish you from other biotech companies.

An important thing is for the values to be specific. Well, all biotech companies should be patient-centric, teamwork is always important, honesty (well, who wants to be dishonest?) but what is it specifically that makes let's say ABC Therapeutics... ABC Therapeutics? What is ABC's unique DNA? ABC has let's say 100 employees today. The values are there already! ABC's DNA has been formed; that's why it would be worth asking everyone and putting ABC under the microscope to identify its unique cultural DNA... that no other biotech out there has!

So what you want ideally is someone Googling the ABC values and getting back... the ABC website! :) You talk to an investor about your values and they go "Oh, you're right... that's soooo ABC!"

3. The values have to be actionable and they must mean the same thing to everyone. HR often initiates this kind of exercise and I asked HR in the past... so what do you think the values should be? And I get answers like *innovation, patient-centric, integrity, respect, honesty...* These are all great but what does *integrity* mean? If we ask the 70 people at ABC, we will get a different definition of integrity... We do want to capture all that input.

For example, an example of integrity and honesty from my experience is ... we are not going to sign a deal with pharma just to sign a deal and drive the share price up unless this is a clearly long-term thing to do for both us and the partner. We can say no to things and set clear expectations right from the start of the partnership, to create a win-win and protect both ourselves, the partner, and obviously, our vision and patients at the end of the day.

[2 and 3 above can help you craft your message accordingly when collecting input from everyone. For example, you can purposefully decide to not mention the word values in your message and just ask questions... Who are we as a company? What differentiates us from other biotech companies? What do we stand for? What are we never willing to do as a company? etc.]

THE BLACK-AND-WHITE SWITCH

This is one of the biggest realisations from my work with biotech leaders...

Your job becomes 100x easier, simpler, more effective, more enjoyable—and less stressful and draining—when you start making decisions like this:

Is this going to help the patients?

Does this serve the long-term vision of my company and what the hell we set out to achieve here?

No. Then, I don't do it. Period.

But how about the markets? The share price? The analysts? The press release? The current raise? This potential partner? My relationship with this person?

That's all NOISE.

But what if I die in the short term?

You will die anyway!

And I get it, it's not easy, I empathise with you, you get pulled in all directions, your emotions mess up with you...

It's not always black and white, but there are times when you should make it black and white!

You will all of a sudden breathe again, you will feel lighter, more confident, and more in control —and you will win!

Paradoxically, those times are often some of the most critical times when you think you shouldn't go black and white.

I've seen leaders who find themselves stuck in this full-colour conundrum, and the moment they turn on that..... what do I call it... maybe... "the subtle art of not giving a f*ck black-and-white" switch... everything falls into place. Oxygen! Aaaahhhhh...

And they may be still in the same deep sh*t, but boy, are they different. You can sense their confidence; they source incredible power from their decision to do the tough, but right thing for the company.

And when they think that everything is lost, they win.

And they turn crises into the biggest opportunities to grow and create culture. And morale goes through the roof, and they earn the trust of the whole company, their board, investors, and everyone wants to follow them—including myself who wants to crash into my Zoom screen and kiss them.

I'm the luckiest person in the world to have worked with those leaders and witnessed the miracles they created.

It's been an extremely difficult environment for biotech companies lately... and the best time to show your leadership, build a great company, and win.

HOW CAN THE TEAM SEE THE ELEPHANT?

Whether you are a startup team of 3 or a Google team of 100K people, each one of the 3 or 100K sees something that the other 2 or 99999 don't see. Each one has some sort of micro-information that is not available to the others. Each one can see a part of the elephant, but not the elephant.

So if we assume a clear and inspiring mission, the biggest leverage you have towards achieving that mission is the dialogue, i.e. how well the players communicate with each other about the available options and trade-offs with respect to the mission.

"How do you see this? This is what I see. What if we did this? What is going to happen? What do you think?"

It's time-consuming but I personally don't know any other way… The players have to communicate—and do it well.

And to do it well, they have to want the collective good and be willing to check their ego at the door. They have to be able to evaluate one strategy over the other in a fact-driven, mission-first and "ego-proof" way.

And to achieve this, you need a culture of trust. You need to invest a whole lot of resources and conscious effort to help the players trust each other. You need meetings, 1-on-1's, training, mentoring, outdoor activities, getting-to-know-each-other's … and I don't know what else you'll need!

There are no shortcuts, or let me say, this is the most slowing-down and at the same time lightning-speed shortcut towards company success that I know.

MY REFLECTIONS ON JOHN MARAGANORE'S REFLECTIONS ON ALNYLAM

I've read John Maraganore's essay "Reflections on Alnylam" a few times and I recommend it to all my CEOs.

Here are 12 key lessons from building and leading an iconic biotech company.

1. TRUST THAT VOICE.

If you can't stop thinking about it, you have to do it...

"Driving into work each morning, I couldn't stop thinking about the potential of RNAi as a new approach. Yet many friends and colleagues (including my two older children) thought I was making a mistake and cautioned against the move. After the 'genomics bubble' burst, the biotech sector was in a dark winter, and investment in novel science was highly disfavored. Millennium CEO Mark Levin who was a friend and a mentor, encouraged me to pursue what excited me most. What I kept coming back to was that if the technology hurdles could be conquered, RNAi therapeutics were too great of an opportunity to walk away from."

2. HAVE A CLEAR VISION AND MISSION.

Pay attention to the subtle distinction here: The vision is the dream, your true north. It inspires you and keeps you going. The mission is what the company does and how it's going to achieve its vision.

"At my first board meeting as Alnylam CEO in December 2002, I presented my vision and mission to crystallize the company's aspirations and direction. This was something Levin had taught me at Millennium. Our vision: Harness a revolution in biology for human health. Our mission: Build an independent, top-tier biopharmaceutical company founded on RNAi. Remarkably, those official words have not changed since."

3. ALWAYS STAY LEAN AND HUNGRY.

Constraints can spark creativity, resourcefulness and drive. The Alnylam team had no other option but to make it work—for patients and themselves! John's focus on independence, urgency and impact all come together here:

"These remarkably different outcomes underscore the power of culture in biotech. At Alnylam, we were willing to take appropriate risks, advancing even 'imperfect' molecules into development to safely learn from early human studies. Furthermore, as a focused pure play, Alnylam had a 'fear of mortality' that made it essential for us to succeed in bringing RNAi therapeutics to market. But Sirna was part of a larger company, needing to fulfill certain criteria around drug candidates, and RNAi was very far from a 'life or death' proposition. The acclaimed management consultant Peter Drucker once claimed that "culture eats strategy for breakfast." In the case of Alnylam's story—and perhaps many biotechs—that was an understatement!"

4. IN BIOTECH, YOU IPO TO BUILD, NOT TO CELEBRATE.

John: "I knew from the beginning that Alnylam would need to raise substantial capital from both investors and pharmaceutical partners to bring RNAi therapeutics to patients."

The above explains the below:

"IPOs are often romanticized by company management and boards as a special 'Kitty Hawk' moment, but in reality they are simply financing events. To us, it was key that we would have sufficient news flow with science and pipeline progress, and partnerships to garner continued interest by public investors. We were confident that we would."

5. PHARMA, ALL I NEED IS YOUR MONEY AND YOUR NAME.

This is what a pharma alliance is: Dear pharma, give me your money and your name, so I can get more money to get this thing to the patient and realise my vision. And don't worry, this will be a win-win. I love you, Biotech.

Alnylam also benefited from partnerships with major pharmaceutical companies. Having watched Millennium's remarkable deal-making over the prior decade, I recognized that pharma alliances were mostly about funding and external validation and, if structured correctly, could be associated with a minimal 'tax' on the company's abilities to advance its own science and pipeline. Although there are notable exceptions, rarely do partnerships provide the 'big brother' benefits (such as drug discovery and development expertise) often advertised.

6. DURING YOUR 15+ YEAR JOURNEY MANY WILL LOSE FAITH IN YOU.

Just soldier on and "keep a copy of that article by your desk"...

"The RNAi therapeutics downturn began in September 2010, when Novartis declined to exercise its $100 million option to acquire broad nonexclusive rights to Alnylam's technology. After this, a combination of leadership changes and recession-driven profit-and-loss pressures at Roche led it to jettison its RNAi investment after just a three-year, toe-dipping sojourn. To say the least, the external sentiment about RNAi turned sharply sour. In early 2011, New York Times science reporter Andrew Pollack described it best in an article titled "Drugmakers' Fever for the Power of RNA Interference Has Cooled." I keep a paper copy of this article by my desk to this day."

7. IF YOU EVER HAVE TO DO THE MOST PAINFUL THING IN THE WORLD, DO IT WITH EMPATHY AND COMPASSION.

It was not their fault. You know that. It was not your fault. They know that. All you have to do is take care of your most valuable asset: your people.

"We had no choice but to reduce our workforce so that we could 'live another day'. In one of the most painful decisions of my career as CEO, we let ~25% of our workforce go in September 2010 and then another ~33% at the start of 2012. I learned the importance of dealing with a tough moment like this with empathy and compassion. Barry and I actively worked to find new roles for all our departing employees. We kept a list of affected employees and their new potential job prospects, and reviewed it weekly at our management board meeting. I personally reached out to many contacts across the industry to find homes for our departing people."

8. ALWAYS TRUST YOUR PEOPLE'S JUDGEMENT, ESPECIALLY WHEN THE WORLD IS ON FIRE.

When you're working on the frontiers of science, no manager has the answer. Or if someone does, that will be probably your amazing people who are working close to the problem.

"This showed some early promise, but we were again hampered by a lack of potency, and the longstanding investment in conjugates was wearing thin. In a memorable meeting in my office, Mano appealed for "one last experiment" to evaluate greater stabilization of the siRNA backbone as a way to achieve enhanced potency. Akshay was equally vocal about recognizing the potential for conjugates and having a healthy respect for the challenges of developing an intravenous LNP-based delivery platform. I had learned over the years to listen to my colleagues; after all, we were on the frontiers of science together, and no one had all the answers. I consented to continue the GalNAc effort and realize this last experiment's conclusion. The bet paid off!"

9. FREE UP YOUR PEOPLE AND LET THEM PLAY.

When your company's vision is to "harness a revolution in human health", you need innovation, i.e. kainotomia, i.e. "a new/fresh cutting". Let your team cut away!

Aristophanes first used kainotomia, the Greek word for innovation, in 420 BCE in The Wasps. As a satirist, he mocked innovators as being unusual members of society. This resonated with Alnylam, especially in the early days, as many people questioned the likelihood of our succeeding with RNAi therapeutics. The spirit of kainotomia was our rallying cry to encourage creativity by our scientists and clinicians, but we also applied it across disciplines. We created a '20% time rule' (something I had learned from my days at Biogen in the 1980s), encouraging our scientists to devote as much as one-fifth of their time to pursuing their own ideas. We explicitly discussed kainotomia as one of the key principles in our core value of 'Innovation and Discovery'.

10. EMPATHISE WITH STAKEHOLDERS AND WIN THEM OVER AGAIN.

Investors and partners have fears just like everyone else. If they've lost faith in you, do whatever you have to do to ease their fears.

"I convened members of my core team, including Barry, Akshay and Sara Nochur, our head of regulatory affairs, to discuss a shift of focus from platform to pipeline. I was convinced that the only way to restore confidence in RNAi was to demonstrate unassailable human POC results. Using the analogy of hearing jingle bells and believing in Santa Claus (from the children's story The Polar Express), I reasoned that our external stakeholders needed to hear the 'bells' of human data."

11. MARCH YOUR TEAM WITH A CLEAR AND FOCUSED STRATEGY.

The 5x15 strategy helped everyone push in the same direction during a critical time for the company. And it didn't lack ambition. A 2x15 or 3x15 strategy would have sent the wrong signal: "we're aiming for survival..."

"We discussed communicating a set of five-year goals at the upcoming January 2011 J.P. Morgan conference. I wanted to propose a new strategy called 'Alnylam 5×15', with a commitment to advance five RNAi therapeutic programs into clinical development by the end of 2015. [...] The team pushed back on this. With only one liver-targeting program in development at that time (ALN-TTR01), and no human POC data in hand, the team suggested that '2×15' or '3×15' might be a more manageable goal to promote publicly. I disagreed, saying that a reduced number of targets would not excite our stakeholders nor motivate our team. And, so, Alnylam 5×15 it was! Our research organization rallied behind the new strategy."

12. ALWAYS BE TRANSPARENT.

This is not just about protecting the trust of your partners. No, no, no. This follows directly from the vision. How can a patient-centric company, embarking on a journey to "harness a revolution in biology for human health", not be transparent?

"While ENDEAVOR was enrolling, we heard reports of worsening neuropathy in the ongoing revusiran phase 2 open-label study [...] Out of the abundance of caution, we asked the independent Data and Safety Monitoring Board (DSMB) of the ENDEAVOR study to conduct an unblinded assessment [...] The DSMB had recommended that we stop ENDEAVOR, not due to a neuropathy finding, but due

to an imbalance of mortality against the drug arm. We immediately moved to discontinue the study. Having just hired Yvonne Greenstreet out of big pharma as our COO, I asked her to coordinate our efforts in ensuring effective communication with our many stakeholders. [...] We launched an internal investigation to learn more [...] We achieved a remarkable level of transparency throughout this trying and uncertain period."

TAKEAWAYS FROM THE BOOK "FROM BREAKTHROUGH TO BLOCKBUSTER"

I've discovered a hidden gem! From Breakthrough to Blockbuster—The Business of Biotech. A book about the players, the numbers, and the challenges in the biotech industry.

Here are 10 key insights for biotech leaders:

1. MOLECULES + MONEY = MEDICINES

The scientists need the financiers; the financiers need the scientists. It's a marriage, a 100% symbiotic relationship. To create new medicines, both sides must be able to see the world through each other's eyes and work well together.

"Biotech companies are the offspring of a marriage of convenience between science and money. These two essential components are jointly invested in the pursuit of profits and promising medicines."

2. SEVEN OF THE TOP TEN BEST-SELLING DRUGS IN 2019 CAME FROM BIOTECH COMPANIES.

No matter the hurdles, biotech companies have clearly disrupted the drug development market...

"An analysis of all products approved by the Food and Drug Administration (FDA) from 1998 to 2016 indicates that biotech companies created nearly 40% more of the most important medicines for unmet medical needs and did so with much lower overall costs. These breakthrough products have often become blockbusters as well. Most of the ten best-selling medicines in 2019 originated in biotech companies."

3. PERSUASION IS THE MOST IMPORTANT SCIENCE.

And your persuasive edge will come from deeply understanding your partners:

- agenda, goals, motivations
- mindset
- concerns, fears
- internal structures, decision-makers, politics

"One of the key reasons for identifying these influential and important groups is to emphasize the need for entrepreneurs to be able to speak convincingly to each in their own language. As entrepreneurs deal with these groups, they will rarely, if ever, have the upper hand. Accordingly, they will need to manage the relationships primarily by articulating persuasive arguments showing that if the stakeholders can help the company fulfill its mission, the stakeholders will, at the same time, be successfully achieving their own goals."

4. IT DOESN'T COST $1.4 BILLION; IT COSTS $2.56 BILLION IF YOU INCLUDE TIME COSTS!

There's a lot of debate on what you should include or exclude when you calculate drug development costs but no one would argue with Einstein that compound interest is the 8th wonder of the world!

"If all the money that is spent in drug development were instead invested in stocks and bonds, the investor would have a lot more money at the end of a dozen years than at the beginning because of the financial miracle of compound interest. For example, money invested at a 10% compounding interest rate will double just about every five years."

The average time from preclinical research through FDA approval is 15 years. Why would an investor tie up their money with you and not put it in something else? 15 years can do financial wonders. Think carefully about potential acquisition/exit points and where those fit in along your trajectory to keep selling your story to investors.

5. IF ONLY A DRUG WOULD COST ONLY $2.56 BILLION...

Well, the following puts things into further perspective...

"While the lost opportunity costs essentially double the overall costs of drug development, another factor plays an even more dramatic role. The single largest cost associated with the development of any successful pharmaceutical product is the cost of all the unsuccessful drugs that had to be discovered, developed, and tested to find the one that finally works."

6. DRUG DEVELOPMENT COSTS ARE CONSTANTLY RISING.

Forget Einstein for now. If we look at the "real", out-of-pocket drug development costs, those have grown from $0.52 billion in 2003 to $1.4 billion in 2013 (both figures in 2013 dollars). That's a 166% increase in a decade!

Drug development costs are going up mainly because clinical failure rates are also going up. The chances of getting FDA approval in 2003 were 20% whereas in 2013 only...12%. And here are some reasons for that:

"Researchers and industry analysts have offered a number of reasons that could explain this drop: the "better than the Beatles" problem, where new drugs are designed to be not only as good as but also better than the best current standard of treatment; the focus on more challenging targets and diseases; the increasing complexity of clinical trials and the hurdles for regulatory approval as well as reimbursement pressures (for example, demonstrating benefit with respect to comparator drugs rather than placebos)."

7. YOUR JOB IS TO KEEP SELLING YOUR STOCK!

Back to the art of persuasion… I loved this quote by my fellow Greek Stelios Papadopoulos, aka the "Godfather" of biotech:

"One veteran investment banker, Stelios Papadopoulos, said to a meeting of biotech executives one of us attended, 'Until you get a product on the market, all you've got to sell is stock.' No matter how exciting the company's scientific work may be, the money needed for those research and development efforts has been as important to the company's overall success as the molecules being developed."

8. NO PARTY HAS A SECRET ADVANTAGE IN BIOTECH-PHARMA ALLIANCES.

Deep knowledge of technology does not provide better insights into the ultimate value of the product. No one can better predict which new treatments will make it through the Valley of Death.

"In something as unpredictable as drug development, there is rarely a chance to have a secret advantage. There are so many unknowns in the development process that neither buyer nor seller is likely to have an asymmetrical edge. This inability of all parties to predict the future commercial value of a potential drug accurately is not limited to early-stage programs. It can be seen even in the licensing of products that have completed nearly all stages of development."

9. PHARMA IS SLOW... ADJUST EXPECTATIONS.

It takes pharma a lot longer to reach an authoritative decision than the biotech company expects, even on matters where both companies are fully aligned.

"For the biotech company, the partnered product may be the crown jewel, but within a large pharmaceutical company that is managing dozens or even hundreds of ongoing R&D projects, many decisions about this particular alliance require the approval of committees responsible for evaluating all individual research programs in light of the company's numerous other opportunities and obligations."

[Btw, this applies to the early days too. You should be developing relationships with pharma from Day #1. No responses or initial interest says nothing about the value of your technology. They're slow. They're busy. Something else has their attention. Persevere.]

10. THE FUTURE OF BIOTECH IS IN THE HANDS OF PUBLIC AND PRIVATE HEALTHCARE PAYERS.

If governments and insurance companies can't pay premium prices for drugs, institutional investors won't invest in VC funds, which means there'll be no VC funding, no biotechs, and no new drugs.

"To sustain venture investing in biotechnology, the venture capitalists, and especially their investors, need to be confident that the people making decisions at the pharmaceutical companies or investment funds that make those exits possible will continue to believe two things: first, that a product that beats these daunting odds will be approved for commercial sale by the regulatory agencies in the most important pharmaceutical markets; and second, that such an approval will create sales of the product at high enough profit levels to provide an attractive return on the very substantial investment required to pay for this successful drug development effort plus pay for all of the unsuccessful ones. If these beliefs no longer seem reasonable, then the flow of money pouring into biomedical research will screech to a halt, and the associated economic growth will stop along with it."

To me, that last point was fascinating and humbling at the same time, because it unites everyone in biotech, especially after August 2022 when the Inflation Reduction Act (IRA) was signed. Because although everyone has their own agenda, everyone wants the biotech industry to thrive and counts on policymakers for that.

LESSONS FROM THE BOOK FOR BLOOD AND MONEY BY NATHAN VARDI

It's easier and more fun to learn through stories. And here's a captivating story about the long Odyssey of a biotech company that made it to Ithaca.

Pharmacyclics, founded in 1991, is a biopharmaceutical company that gained prominence for its work in oncology, particularly in the development of novel targeted therapies for cancer treatment. Under the leadership of its CEO, Robert Duggan, who took the helm in 2008, the company reached new heights of innovation and success. Its flagship drug, Imbruvica, was a game-changer in the field of hematologic malignancies, revolutionizing the treatment of conditions like chronic lymphocytic leukemia and mantle cell lymphoma. Duggan's strategic acumen and unwavering commitment to advancing therapeutics led to the company's acquisition by AbbVie in 2015.

Here are eight insights for those who want to succeed and have an impact in this beautifully unique business:

1. AMBITION + PURPOSE > SKILL.

Bob Duggan, the man who took over Pharmacyclics from its founder and Stanford professor Richard Miller, was neither a scientist nor a doctor. He was a college dropout!

"Robert W. Duggan was not destined to lead a biotechnology company. He had no scientific training and no experience in the highly regulated biopharma industry. Zero. The companies in the sector were generally led by gray-haired men, often with MDs or other advanced degrees, who had climbed corporate and academic ladders for decades. Duggan didn't even have a college diploma. But at the age of fifty-two, what Duggan had was a track record of business success and supreme confidence in himself. He literally believed he could accomplish anything if he put his mind to it."

Yet, he had a strong why. After the tragic death of his son at age 26 from cancer, Duggan had an unwavering determination to save lives:

"But even with financial markets and the global economy on the precipice, Duggan found purpose in the company. He wanted to find a medicine that would change people's lives, especially those with brain cancer. 'We take full responsibility for what we know and we know the work we are engaged in is work worth doing.' he said at the time about his decision."

Maky Zanganeh, the Head of Business Development who later became Pharmacyclics' COO and the second most powerful person in the company, had no pharma or oncology experience either. She was a ... dentist!

"Duggan would soon promote Zanganeh to the position of Chief Operating Officer. But even before he did, it was clear that Zanganeh had become the second-most powerful person at Pharmacyclics— the person Duggan valued above all others. Exceptionally smart and a shrewd negotiator, Zanganeh had Duggan's complete trust. Together, they challenged long-held industry assumptions and made sure to keep in close contact with the investment community—people like the Baker brothers, Felix and Julian, whose hedge fund had become Pharmacyclics' second-biggest shareholder. Both Duggan and Zanganeh worked extremely hard—they lived and breathed Pharmacyclics."

2. AN OUTSIDER CHALLENGES BELIEFS AND PUSHES FOR INNOVATION.

If you're not an outsider, surround yourself with people who're going to challenge your assumptions.

"Duggan questioned things. Why couldn't they get a discount on the CT scans they were ordering for patient trials? The life sciences veterans had to carefully explain that these tactics were industry no-nos, as there could not be bias in the decisio of whether to enroll a

patient in a clinical trial. Duggan and Zanganeh mostly accepted the pushback, but kept prodding people to be innovative."

3. GOD, SAVE THE BILLIONAIRES.

Duggan loaned Pharmacyclics $6.4 million in 2009 when investors were running for the hills.

Without his loan, the company would have gone bankrupt and never developed the drug that's saving hundreds of thousands of blood cancer patients today.

"But Miller had left, and Duggan was holding the bag of a cratering company running dangerously low on cash. Days after Miller's exit, Lehman Brothers, the big New York investment bank, collapsed. The financial crisis left Pharmacyclics with few options. Raising money had become impossible. Investors were running for the hills. Many small biotech companies had been forced to restructure, and some filed for bankruptcy. Pharmacyclics seemed headed in that direction as well."

Wayne Rothbaum, the man who started the rivalry company that developed the other blockbuster drug, was a billionaire New York stock trader. He invested one-third of his net worth into the company and convinced his billionaire friends to shoulder the rest of the investment.

"At one point, Dave Johnson called up Rothbaum and told him the company coffers had only six weeks of cash left. Rothbaum raged over the company's inability to accurately budget. He had been told a few weeks earlier that Acerta had enough cash to get through the year. Now it was almost bankrupt. Rothbaum was forced to accelerate Acerta's next phase of fundraising, a Series B round planned for the late summer. In less than three weeks, Rothbaum pulled it off, and Acerta had raised $3 75 million."

You can't start a biotech company in your garage. It costs more than $1 billion to bring a drug to market. I agree with the author that we need venture capitalists, hedge fund managers and Wall Street to be involved in drug research.

Also, these guys are not just "backers". They have insight, they're extremely knowledgeable, they've seen patterns and they're active players in the fight against cancer.

"Rothbaum's entire identity had become wrapped up in Acerta. He took a step back from trading stocks and mostly focused on the company. For years, he had been an interested spectator of biotechnology companies. For the first time, Rothbaum was now deeply involved. To speed things up, he wanted to make sure Acerta did things in parallel and not sequentially. From the preclinical work in the laboratory to tax strategies, Rothbaum had a hand in almost everything. Sometimes his obsessiveness yielded important results."

4. THE REGULATORS ARE PEOPLE.

You're not dealing with the FDA; you're dealing with people. Believing that regulators exist to make your life harder won't get you far. Richard Miller, the founder and ex-CEO of Pharmacyclics had declared war against the FDA. It didn't help:

Miller sketched out an opinion article that blasted the FDA for denying treatments to cancer patients. The gap between medicine and statistics had simply paralyzed the FDA approval process, he wrote. "The FDA bases its approvals—for everything from medications for minor ailments to new cancer treatments—on the rigid application of the same outdated statistical standards," he expounded. Miller submitted his article to the opinion pages of the

Wall Street Journal. [...] Miller was escalating things further with the FDA, the government regulator that had the power to make or break his company and its experimental drugs, by taking his beef to the pages of the Wall Street Journal . While Miller's argument had merit, many in the biotechnology community thought that publicly contesting the FDA's decision over Xcytrin was madness. How would Pharmacyclics be treated by FDA officials going forward? [...] In August 2007, he wrote a second opinion piece for the Wall Street Journal, arguing that "current FDA policies are discouraging the development of groundbreaking treatments for cancer and other killer diseases." Before the year was out, Miller wrote a third Wall Street Journal opinion piece, decrying the FDA's "cumbersome and overly restrictive policies." He at least expected the FDA to give him a shot. Instead, the FDA sent him a "refuse to file" letter. The regulators were not even going to consider Xcytrin's application and denied the drug without a review. It was strike three for Xcytrin.

When Robert Duggan took up the reins of the company, things changed...

"Rich Pazdur was known in the FDA as Doctor No. It was very hard to get drugs approved by him. But after his wife lost the battle against cancer, Pazdur changed and set out on what he described as "a jihad to streamline the review process and get things out the door faster". Though nominally on opposite sides of the table, Duggan and Pazdur found themselves as allies. Cancer had profoundly affected their lives on a personal level. The disease had changed them both. In fact, the reason Duggan and Pazdur were in a room together talking about ibrutinib was that Pazdur's office had repeatedly rejected the brain cancer drug Richard Miller had championed, leading to Duggan's takeover of Pharmacyclics."

Regulators are walking on a very thin line. Being a little too loose or being a little too tight both risk the lives of thousands of patients.

"In the end, the FDA is in an impossible situation, and I have been in that impossible situation. You are either approving drugs too fast, you are approving drugs too slow," is how Pazdur would explain it. *"But what we try to do is establish a balance of safety and efficacy."*

5. CAPITAL WILL ALWAYS BE AVAILABLE TO THOSE WHO WANT TO REDEFINE HUMAN HEALTH.

The current economic environment may look gloomy for biotech and all industries (February 2023), but I've always held a firm belief that capital will always be available to the crazy ones, the misfits, the rebels...

"Even as the Great Recession hit the economy, Duggan did not believe there was ever a scarcity of money or ideas. There was only a scarcity of confidence in the ideas. The key to building confidence was to do whatever you say you are going to do and not promise to do what you can't accomplish."

6. BIOTECH IS A PEOPLE'S GAME.

Vardi said in an interview: "In my book, I wanted to answer one question: how amazing drugs are created today."

To me, Vardi's book is about people: human emotions, egos, motivation, ambition and the transcendence of human existence. To bring a breakthrough to the patient, extraordinary effort is required from scientists, biopharma, investors, and regulators. All these players have their own legitimate and honourable agenda. There's no way you can succeed in this business unless you've understood each player's motivations. And you may do all this and still fail, which brings me to the last point...

7. YOU ALSO HAVE TO BE LUCKY.

Oh yes, For Blood and Money is about luck too. It's extremely humbling.

"From personal experience, Hamdy understood that developing cancer drugs was like buying a lottery ticket. Most cancer drug researchers were stumbling around in the dark. Occasionally, a variety of factors—including luck—came together to produce a winning treatment. It was a game of overwhelmingly negative odds. The vast majority of novel cancer drugs tested in patients failed."

8. EVERY MOTIVATION IS BEAUTIFUL.

Although the book is called For Blood and Money, I don't think that biotech is the easiest path to riches. Every character in this book, including the hedge fund managers, is motivated by something beyond money. In my case, seeing founders/CEOs who were willing to go against the odds to get their technology to the patient was seductive! I thought, if I can support these guys and increase their odds of success, I'm having my own impact on human health.

But that's just my motivation. To me, every motivation is beautiful and necessary in this business as long as one is not being unethical. So, for blood... or money.

THE MOST IMPORTANT MINDSET SHIFT FOR SCIENTISTS WHO COME OUT OF ACADEMIA TO BUILD A BIOTECH COMPANY

The traditional way to spin a company out of academia is.. you do all this hard work, publish a paper, and then you think "oh maybe there's a company here!". But the challenge with this approach is that you might do years of work and realise that you've solved the wrong problem.

You may think that partners care about scientific ideas. They don't. VCs will invest in you if they think you can create a billion-dollar company. Pharma will partner with you if you can help them achieve their goals. And they're not evil, everyone just has their own goals.

In healthcare, it's not about how you can use your technology to help the patient. It's about using whatever technology you have to use to help the patient. In this end, you may have to use technologies that you never used in your research career.

Academia and industry are two different worlds. George Church has said, "As an academic, your job is to spend money, which means you can take gigantic leaps... almost artistic leaps! In corporate, you have to make money, which means you can't take those gigantic leaps".

So how do you know which leaps to take? Talk to those who'll be the users:

"Let's assume I can solve this. Does that make you excited? Would you pay for it? What else would you need to know before you can use this technology?"

Talk to folks in other companies, Pharma, or investors who've invested in this kind of companies to understand the marketplace and the key challenges.

When you've derisked the market, then it's a matter of "can we actually do it?".

Plus, when you know that there is a market for what you're building, you're more confident, focused and motivated.

Now... all this is easier said than done.

Because in every great scientist, often, there are inner forces that work against this new mindset...

You guys are artists taking artistic leaps as George Church said. And as an artist is obsessed with her art, you're obsessed with your science. That's wonderful! To me, obsession is a form of love. You're obsessed with science, I'm obsessed with my leadership tools and theories. I get you.

The artistic part of yourself may not feel like "commercialising" or talking to partners and investors who can't—and will never—

fully appreciate your art. But, there's a possibility here: to find beauty in making something out of your art and giving it to the patient.

Achieving that won't be easy. Some people will question your art. Others will say you can't do it. You'll hit walls with partners and regulators. You'll question yourself, "why would I ever want to go through all this?"

But it's worth it. It's why you started this company. Because you're now obsessed with the vision, with the impact, with the patient.

Dalton Caldwell, a Partner at YCombinator, has said, "Do you know what gives you passion? The damn thing working. Users. Revenue. Numbers. What you didn't have an idea about and all of a sudden appears to be a business that is working."

Although he was not talking about biotech, this applies to biotech too. It's no longer about just cool scientific ideas. It's about getting faster to the patient and saving lives. That's the only thing that matters. That's the absolute passion and reward.

FUNDRAISING ADVICE—THESE 10 BIOTECH FOUNDERS HAVE RAISED MORE THAN $350 MILLION IN VC FUNDING

1. TELL A STORY THAT SPEAKS TO A UNIQUE OPPORTUNITY.

Investors look for opportunities where they can see the value where no one can. If they think that you're one of the 100 people doing the same thing, how do they know that *you* will be the winner? Tell your story in a way that helps them see why you're doing something different so that they can come in early to support you and make the outsized returns they're after. — Eric Kelsic, CEO at Dyno Thereapeutics

"Your job is to convince the investors that your startup is a better bet than the hundreds that they will turn down this year". — Bryan Mazlish, CEO at Surf Bio

2. YOU WON'T BE A FIT FOR EVERYONE.

"Investors often don't pass because you have a terrible idea; they'll pass because you're not a fit for each other. Some passes are always going to be passes, so try not to let it become demoralizing." — Nick Goldner, CEO at Resistance Bio

3. KEEP DOORS OPEN AND DON'T COMMIT RIGHT AWAY.

"If you're in diligence with multiple investors and one of them gives you a term sheet, use the Fear of Missing Out to get as many term sheets as you can while telling the first investor that you're considering their offer. You don't need to accept that first offer, and you definitely don't need to accept it right away. You need to do what's best for your company, that is, building as much value as possible by increasing optionality. Investors respect this." — Cheri Ackerman, CEO at Concerto Biosciences

4. BUILD THE RELATIONSHIP BEFORE YOU NEED THE MONEY.

Be a stage ahead. If an investor can follow your development for a couple of years, it makes it a lot easier to raise the money when the time comes. — Chris Gardner, CEO at Sequence Bio

It's never too early to start building connections with VCs. Ensuring that they have a longitudinal and positive evaluation of you as a person before raising money from them can be incredibly beneficial. — Joshua Young Yang, CEO at Glyphic Biotechnologies

5. INVESTORS WANT $$$, NOT COOL SCIENCE.

VCs won't fund your cool scientific idea. VCs will fund you if they think you can turn your idea into a billion-dollar company. — David Li, CEO at Meliora Therapeutics

6. USE EVERY INTERACTION AS AN OPPORTUNITY TO LEARN.

Even if you don't come out of a partner pitch with a term sheet you will leave with valuable learnings that you can use to strengthen the business. — James Field, CEO at LabGenius

7. "FINANCING IS A MEANS TO AN END, NOT THE GOAL ITSELF.

The goal is to build an enduring, useful company for customers. Think about what's needed to build that kind of company, and backsolve into the amount of money required to reach that goal." — Ramji Srinivasan, CEO at Teiko Bio

8. VET THEM AS MUCH AS THEY'RE VETTING YOU.

"When talking to investors, your goal should be to create a relatively even information exchange. Ask them how their firm supports companies like yours and how they relate to founders. We had set a goal to have term sheets by the end of the summer, and we told each investor that.

This gives you the opportunity to see who is respecting your time and who is dragging you along." — Cheri Ackerman, CEO at Concerto Biosciences

9. IT'S NOT JUST ABOUT THE SCIENCE; IT'S ABOUT YOU.

Be yourself and be willing to share your story, aspirations, and motivations, and not just focus on the science. I learned that early-stage investors are betting on you and your team. — Nabiha Saklayen at Cellino

My $0.02: There are 100s of reasons a VC won't invest in you that you're completely unaware of:

- bad memories of deals gone south
- internal politics
- conflicts of interest with other portfolio companies
- an investor's unconscious fears and biases

Yet, founders will come up with 100s of reasons to make it ALL about themselves.

I see founders who want to change everything in their pitch after talking to only one investor—and without ever having received any direct feedback from them! And I go... how do you know???

There's as much randomness in the fundraising process as there is in people!

How you deal with randomness:

- ask for direct feedback
- don't change your hypotheses unless you have a sufficient statistical sample
- kiss a lot of frogs to find the prince

Oh, and all this randomness applies not only to an investor's nos ... but to their yeses too! You may think that it's your bulletproof technology that got you funded... but is it?

Remember that investors are always taking bets... especially in an industry with 10-year lifecycles and 98% failure rates in clinical trials...

In the end, what will get you funded is your passion, your perseverance, your commitment to solving the problem you've decided to solve.

It's your belief in your vision that will increase the investors' belief in you.

WHAT I LEARNED FROM READING 70+ INTERVIEWS WITH BIOTECH FOUNDERS

I spent over 100 hours researching and reflecting on 70+ interviews with biotech founders. Here are 7 principles these founders all agree on:

1. FALL IN LOVE WITH THE PROBLEM, NOT YOUR TECHNOLOGY.

Building in healthcare is not about how you can use your technology to help the patient. It's about using whatever technology you have to use to solve the patient's problem. To borrow from Steve Jobs, "start with the patient and work backwards for the technology".

2. FOCUS ON YOUR VALUE PROPOSITION

This applies to both investors and partners. "VCs won't fund your cool scientific idea. VCs will fund you if they think you can turn your idea into a billion-dollar company." — David Li, Founder and CEO at Meliora Therapeutics

"For biotech companies, it takes 5 to 7 years to get through the preclinical and regulatory work, if you are lucky. Why would an investor tie up their money with you when they could put it into an S&P 500 company and see those returns much more quickly?" — Jonathan Thon at Strm Bio

With Pharma, you have to be strategic too. Your solution has to align with their focus and solutions. You have to match their targeted shopping list.

"Talk to folks who'd be the users: Let's assume I can solve this. Does that make you excited? Would you pay for it? What else would you need to know before you can use this solution?" — Eric Kelsic, CEO at Dyno Thereapeutics

3. TALK TO INVESTORS AND PARTNERS TODAY

Don't wait until you have the perfect deck, data, or results in your hands. Talk to investors, partners, and Pharma today. Develop those relationships early.

"Building in biotech is a relationship game. People do deals with people who they like, who they respect, who they want to work with." — Becky Pferdehirt, investor at a16z

4. YOU MUST GIVE INFORMATION TO PARTNERS.

Early-stage biotechs are too worried about IP protection and their ideas being stolen. But how can you expect your partner to trust you if you're not willing to give information?

When you give information:

- you learn faster
- you pressure test your assumptions
- you build credibility
- you strengthen the relationship with all those you'll team up with to get to the patient

5. BE HUMBLE (BOTH AS A LEADER AND AS A COMPANY)

Building in biotech is 100% interdisciplinary team effort within the scope of the company and a fully collaborative effort outside the scope of the company with investors, partners, Pharma, etc.

"Humility is an under-appreciated skill in founders. I know the area where my edge is. In all other areas, there are smarter people who better understand the details of our processes." — Nicolas Tilmans, Founder and CEO at Anagenex

"Listen to your team. They are the ones closest to the problem and the ones with the right insights." — Ramji Srinivasan, CEO at Teiko Bio

I'll never forget what a Sr. Director at a big pharma company said to me: "I appreciate the founders who can be honest and open about the limitations of their technology. No one's solution is perfect including ours."

6. GET THE RIGHT PEOPLE ON THE BUS.

Building in biotech is a long and bumpy ride. Don't whitewash the risks when hiring. Tell your first hires that it's going to be hard but if you succeed you'll save lives and redefine human health.

"Make sure that your people are really bought into your vision and the idea because it's going to be a long-term effort. They must be really excited and jazzed about the idea and willing to go through the hard work and the ups and downs." — Tim Lu, CEO at Senti Biosciences

7. BECOME A STORYTELLER.

Can you tell us your vision without talking about the science? Why does it matter? What is your impact on the patient and human health?

"You have to tell your story with sincere passion and excitement and communicate complex scientific concepts in a simplified and compelling manner adjustable to your listeners." — Diego Rey, Co-founder and CSO at Endpoint Health

"Don't assume too much knowledge on the part of investors. Say, I know our approach is somewhat complex are there any holes for you in what we covered?" — Sara Nayeem, Investor at Avoro Ventures

A 6-STEP FRAMEWORK ON HOW TO MANAGE YOUR BOARD AND INVESTORS

Biotech CEOs and management team, I understand you and empathise with you when you're getting pulled all over the place by your board and investors. Here is my 6-Step Framework towards feeling more empowered and less overwhelmed.

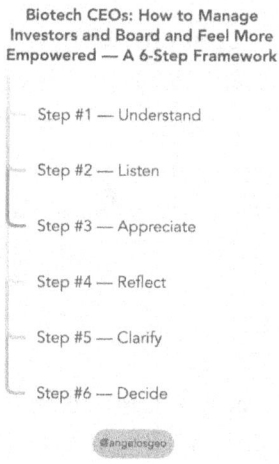

Biotech CEOs: How to Manage
Investors and Board and Feel More
Empowered — A 6-Step Framework

Step #1 — Understand

Step #2 — Listen

Step #3 — Appreciate

Step #4 — Reflect

Step #5 — Clarify

Step #6 — Decide

@angelosgeo

STEP #1 — UNDERSTAND

First, you have to understand that the job of your investors and board is exactly that, i.e. bombard you with hundreds of different—and sometimes contradicting or opposing—opinions.

The more diverse the board, the more diverse their opinions—which is a healthy and desirable thing! A lot of investors/board members will often say, "Don't listen to me!" but that won't stop them from telling you what to do the next day!

Again, that's their job and you're lucky to have active and passionate partners offering perspective and keeping you sharp.

I say partners, because, to me, the investors and the board can be extremely valuable partners to the company. These partners know very well that they're not the ones who run the company and that they don't have access to all the information and operating view that you have. All they want is you to be a strong team and make the right decisions.

STEP #2 — LISTEN

Listen to all different opinions with an open mind. Where is this person coming from? Put aside all your filters, convictions and preconceptions. Ask more questions, learn, and take notes. Remember that your investors and board, like every human being on earth, want to feel heard and valued. They don't expect you to do what they say but they do expect you to at least listen.

STEP #3 — APPRECIATE

Thank them for their advice and tell them that you're going to seriously consider it. Also, tell them that you'll need to think more to come up with a plan. You don't have to commit to anyone or anything right away. They don't want that either!

STEP #4 — REFLECT

Does their advice make sense? Why? Why not? Is there something that you can see that they're not seeing?

Be open to changing your view. It is a strength to show agility of thinking and responsiveness to other views.

STEP #5 — CLARIFY (AND LISTEN)

Reconnect with them and explain how you see the situation. Help them see what you can see that they're probably not seeing. Be open to being offered more perspective once they have an updated view and go back to Step #2 if necessary.

Obviously, you can't go back to the board for every little decision because you'll become inefficient; you have to go at 100 miles per hour! At the same time, you have to be radically transparent. You don't want to keep your investors and board in the dark. When in doubt, share. They will buy in a lot more when they know that there's no secret being kept from you.

STEP #6 — DECIDE

After having considered all options, decide what's best for the company. Tell your investors and the board why you think this is the "best" decision. Be transparent about the risks, potential outcomes and available future options. The best decision will not always turn out the right one! They know that. Ask for their support and create buy-in.

Again, your board wants you to feel autonomous and empowered. Your relationship is a delicate balance of "power" and influence. The board manages you but you also manage the board, i.e. you manage their concerns and fears by showing that you know what you're doing.

Finally, accept that it won't always be easy to stay grounded and help everyone understand and support your decisions. Your job is one of the most difficult jobs, and one of the reasons is that you have so many bosses, i.e. the board, investors, and all other stakeholders!

With that said, take whatever works for you in this framework and bin the rest! You are running the show and your board, investors and myself, all want you to succeed!

BIOTECH FOUNDERS ARE OLYMPIC DECATHLETES!

Here we go…

1. Scientist
2. Drug developer (this is nothing to do with academic science btw)
3. Entrepreneur (product, market, commercialisation)
4. Politician (building and maintaining relationships, networking)
5. Project Manager
6. Psychologist (oh yes, you're the company's psychologist)
7. Salesperson (storyteller, fundraiser, spokesman, communicator)
8. FBI Negotiator (IP, partners, investors, FDA, tech transfer)
9. Team Leader (the coach who must turn a team of stars into a star team)
10. Buddhist monk (master of emotions and resilience playing a 15+ year game against all odds)

What did I miss???

Because you are all the above and everything else I've missed, here's my advice to you: Be an Olympic athlete. Be a killer founder. Be a pro.

You have to grow faster than your company as a person. The world of biotech is so fast-paced and you, the leader on top, have to develop and upscale constantly to keep up.

It takes conscious, deliberate, and intentional practice. Here are 11 practices for biotech Olympians:

1. Surround yourself with the best mentors, coaches, and advisors.
2. Hang out with the smartest people.
3. Talk to other founders who are on a similar journey to yours.
4. Join support communities and networks like Nucleate or the Henri Termeer Foundation
5. Carve time in your calendar every day to read and learn.
6. Read all those great resources on the Pillar website
7. Go for a walk and listen to those amazing podcasts of the BIOS Community, Biotech 2050 and First Rounders
8. Join the Biotech Hangout hosted by Daphne Zohar every Friday
9. Keep up with the latest news in the biotech world (FierceBiotech, The Timmerman Report, Endpoint News, STAT News)
10. Build, but most importantly, maintain those relationships.
11. Take care of your body, mind and spirit.

I was watching this documentary on Netflix about wingsuit flying... and I thought... what are these guys doing? Is this real?

I went back to the office the next day and I thought, biotech success is 100x harder than that!

That's why I fell in love with biotech. That's how you guys won me over. Because I wanted to work with Olympians. Because to me, biotech is an Odyssey with the noblest purpose. So, here's to the Olympians!

WORDS OF WISDOM FOR ASPIRING BIOTECH FOUNDERS COMING OUT OF ACADEMIA

Aspiring Biotech CEOs coming out of academia, here are some words of wisdom from people who have done it before:

"I'm a huge believer that recently minted PhDs—or MDs or MBAs for that matter—who want to go into biotechnology, should work at a more established biotech company...not necessarily pharma because they can be siloed.

If you work at a more established biotech company, like I did in Biogen or others have done at Alnylam or Vertex, you can learn so much from a group of really experienced people.

I think it's really important for an aspiring biotech CEO or leader to get that experience before they automatically plunge into their own entrepreneurial venture. Sometimes you don't need that experience, and that's really cool.

But there's something about getting some scars on your back, learning drug discovery, learning about business development, and doing it with people that are really good.

This experience can really equip future leaders to be better positioned and more successful. That's the advice I always give to recently minted grads who are thinking about a career in biotech…and it doesn't have to be forever. It can be for just for a few years, whatever the case might be. Then you can go off and do your own startup, go to a venture group, or join a smaller company. But I think learning about drug development first is important." — John Maraganore, Former founding CEO at Alnylam

----

For those of you guys who just can't wait and won't listen to John Maraganore, at least, seek mentors like John and learn as much as you can from them!

Find 5-10 people like John, CEOs with clinical, BD, manufacturing, regulatory, commercial expertise and stay close to them!

This is what Robert Clarke, CEO at Kinaset Therapeutics says:

"For the founder-led movement, I cheer you on for following your dream of turning your invention/idea/experiment into something that can help patients in the future. In fact, Kinaset is a founder-led company with three of our experienced C-suite as founders. For less tenured founders, I would only suggest that somewhere early in the genesis of a newco you think about adding a senior team member who does have that longer track record. That experience and their lessons learned can help you achieve your vision." — Robert Clarke, CEO at Kinaset Therapeutics

----

Also, I'll never forget what Chris Garabedian once said to me: "Drug development is not science; it's a whole different thing. Although there's always technical risk, the reason most companies fail is not because of the science, but because they don't do best practices in drug development. The biotech CEO must have the humility to understand that and strive to learn as much as possible from mentors who have been successful in drug development." — Chris Garabedian, CEO at Xontogeny

And then, it's not just drug development, there's your leadership, the people aspect, the culture... That's another big muscle that you must consciously and patiently train that—as all things—requires time and attention.

I personally believe in your ability to learn and become a great biotech CEO. For me, seeing young folks who want to start/lead a biotech company is a celebration. And that's why I LOVE Nucleate and give my time to Nucleate founders.

I believe that the founders who are obsessed with building a great company and creating a real impact for patients will be successful.

Because they will seek mentors, they will check their ego at the door, they will remain humble... and do whatever they have to do to achieve their noble vision.

BIOTECH LEADERS, RECRUIT YOUR "COUNTERBALANCES"

Biotech leaders, don't forget to recruit your "counterbalances" into your team.

In biotech, we talk a lot about "complementary skills", i.e. science vs business, bio vs tech, clinical vs regulatory vs commercial etc. That's only one aspect of diversity in team building.

There's another aspect that's usually overlooked: the diversity of approach/thinking. For example...

If you're a fast decision-maker, you need someone who likes to mull things over.

If you're all about the big picture, you need someone who's all about the details.

If you are the initiator, you need someone who likes to monitor stuff and ensure that things get completed.

Here are some more examples:

- confident vs cautious
- persistent vs pivot-open

- emotional vs unemotional
- insight-driven vs data-driven

Having counterbalances next to you can be uncomfortable or frustrating at times.

You'll drive each other nuts and fight a lot but... you are setting out for an Odyssey... can you survive without this diversity of "being" in your crew?

You'll say... with so much fight... we're going to sink our boat ourselves! No.

If you all share the same vision for the company and are as committed and passionate about the impact you want to create for patients, not only will you be fine but you'll have even more room for productive fight!

Then you'll say... if I wait until I build this perfectly diverse team, I'll probably never leave the port...

Well, you have to build the best team possible. You're not any company. You're selecting the crew for a 15-20 year stormy journey. It's all about the crew...

[Photo: Achilles pays Nestor the price of wisdom, Iliad, Homer. The Achaean leaders gathered to discuss whether or not to abandon the siege of Troy and sail back to Greece. During this meeting, the wise Nestor urges the Achaeans to continue the war, arguing that they have come too far to give up now. He also suggests that they send ambassadors to Achilles, who has been sulking in his tent, to try to persuade him to rejoin the battle. Achilles is a brave and formidable warrior whereas Nestor is a wise and diplomatic mediator...]

LET YOUR SCIENTISTS GO WITH THEIR PASSIONS

When Joshua Boger started Vertex in 1989, the original business plan was to focus on antivirals except HIV.

He recounts, "The reason we said antivirals except HIV was not because we were anti-HIV, but because of the perception within the investment community at that time.

If you mentioned the word virus, they immediately associated you with HIV, the only virus they were familiar with.

Wall Street had decided that HIV companies wouldn't be successful, either scientifically or commercially.

If we had labelled ourselves as an HIV company, we would have struggled to raise money.

In reality, we didn't have any ideas about HIV. However, we had numerous ideas about other viruses. So that's why we said antivirals except HIV.

But guess what... our first product was in HIV!

And that happened because the passionate scientists that you're trying to motivate to do other projects got some ideas about HIV and started to work on them at the midnight to 6 am shift!

I mean this literally, midnight to 6 am, I'm not kidding! That's when they did the work.

And they started getting results... very promising results, and they convinced me that maybe we did have some ideas about HIV!

I've always respected that passion from the scientists..." — Joshua Boger, Founder and former CEO of Vertex Pharmaceuticals

-----

Scientists don't burn out because they work too hard, they burn out when their hard work doesn't go anywhere.

They burn out when they are confused...when nobody tells them how their efforts help the business and have an impact on patients.

They burn out when the leadership doesn't communicate enough with them, i.e. "yesterday they said I should focus on x, today they're saying focus on y... but why???"

They burn out when there is no transparency about change, priorities, strategy, and company transitions.

They burn out when the strong vision they initially bought into is not there anymore.

They burn out when there's no psychological safety at work, i.e. a sense of confidence that management will not embarrass, reject or punish them for speaking up or taking risks.

Leadership is about getting people excited about a North Star and giving them the freedom to figure out how they personally contribute to that mission...

Loving you, Angelos.

ps: "We had an X-ray crystallographer who wanted to be close to the company so he rented the closest department to our Labs which was actually about 100 feet away. Sometimes he didn't go back to his apartment for 3 months. He slept in front of the X-ray generator because it was warm. It ran all the time with 25 KW of power, so at night it was a warm place for him to sleep. We had some pretty dedicated folks..." — Joshua Boger

EVERY BIOTECH CEO SHOULD READ THIS CIA PUBLICATION

Every biotech CEO should read The Psychology of Intelligence Analysis by Richards Heuer.

Because biotech is like international politics. There are so many players and each player has their own agenda for their own reasons. Investors, pharma, regulators, Wall Street, policymakers...

The biggest mistake that a biotech CEO can make is to take things personally.

But, the reality is that no one is doing anything here because they want to harm you, your company, or your vision.

People, governments, and biotech actors are doing the things that serve their own interests.

If Country A does x—and x happens to affect Country B—doesn't mean that Country B did x to harm Country B.

Country B did x because x simply serves Country B!

Or we should better say... Country B did x because Country B *thought* that x would serve Country B best! (Sometimes we overestimate the other player's rationality...)

These are the so-called cause-and-effect cognitive bias explored extensively in this book.

So, everyone has their own agenda as you have yours.

Take nothing personally.

Listen carefully to understand their deeper motivations.

Help them see your point of view and how their actions might unintentionally be affecting the well-being of your company.

This is what we call effective communication or constructive dialogue.

To me, the paragraph below applies to intelligence analysis as much as to biotech fundraising, team communication, Pharma partnerships, FDA conversations.... and everything biotech!

Loving you, Angelos.

ps: This book was part of my reading assignment when I served in the military. It is one of my all-time favorites.

MAKING A DRUG IS NOT ENOUGH

Joshua Boger, the founder of Vertex Pharmaceuticals writes about his mentor Henri Termeer:

"Henri reminded me and challenged me and mentored me in the conviction that making a drug is not enough and building a company is not enough.

Henri foresaw a great threat to patients if mechanisms of innovation were ineffective. That insight inspired a generation of biotech leaders, me among them, to work for both the sustainability and the systemic improvement of the innovation process when building the companies needed to alleviate the burden of human disease.

Henri led with his beliefs and his principles, none more firmly held than the obligation to keep the needs of patients first. His inspirational genius was in the ability to balance proximal concern for the patient today with the long-term benefits of continued innovation for all patients.

This was Henri Termeer's greatest gift. To these challenges he added, with energy and humility and humour, responsibility for the future of medicine." — Joshua Boger, Founder of Vertex Pharmaceuticals

---

So, imagine you are a biotech company that says "We put the patient first". What patient? The patient you're making a drug for? Henri's patient-centric approach goes far beyond that. It's ALL patients, and not only the patients of today but also the patients of the future!

In my work, I call it transcendent leadership. The transcendent leader invites people to join a project that promises to leave a mark in the world that will far transcend the organisation and the lives of those who carried it through. That's why I'm writing this and that's why you recognised the man in the photo.

When you are a transcendent leader, you'll be told that you're out of your mind. Henri was told that multiple times when he wanted to focus on a disease affecting only 5000 patients worldwide and a solution that would require 22,000 placentas per patient per year and would cost $250,000 per patient per year.

But being "Henri crazy" has benefits too:

- You say things like..."The FDA will change. It should change. If this is the right thing for society, if this is the right thing for patients … then there should be a more streamlined basis for approval. We'll just have to get all the people, all the participants in the system lined up ... the patients and the physicians, and explain it to the FDA." (That was back in the early 1980s when the FDA had very rigid approval processes!)

- When your company is almost bankrupt you bring the mother of the one patient who's responded to your treatment in a room full of investors and you convince them to give you their money.

- When you go to present to the FDA Advisory Committee you stand up, you go to the microphone at the end of the meeting, you make your pitch, and basically say "Come on guys, approve this drug", at a meeting at which CEOs don't speak up!

That was Henri...

The transcendent leader inspires, gives hope, creates community, persuades, innovates and ... wins.

I don't know about you but I believe that being "Henri crazy" works. It works in biotech, it works in business, it works in life.

Picture: Young executive Henri Termeer, around 1985, when he became the CEO of Genzyme. From the book Conscience and Courage.

INSPIRATION FROM LEN AND GEORGE AT REGENERON

"Didn't get our first drug approval for 20 years... didn't become profitable for almost 25 years... up until then, financial analysts used to mock us as the definitive example of scientists not knowing how to run a business... as Len says, we became an "overnight success" after 20 years of perceived failure... But over the last 10 years, many say we have the most productive & exciting pipeline in biopharma... Because we were willing to invest in building the right foundation based upon game-changing genetics technologies".

And here are more lessons on being persistent and having a winning strategy...

"What most companies do is... at a very early stage when they really think they're onto something, they go to Pfizer or Amgen or Merck and they sell out, and the company is gone!

They get what seems to be a lot of money upfront, let's say $100m, the couple of guys who started the company make a decent amount of money... but the company is done! It's then all up to the big company, but what happens is ultimately most of these projects die in the big company...

However, our approach is different. Len who pioneered this model said, hey Amgen or Pfizer or Sanofi... if you want to be part of this, you're not gonna get it all, you'll only get 50% of it and we're gonna keep 50% on the back. So if this turns into a billion-dollar drug we've become a big company all of a sudden because we get half of a billion dollar.

But that's not all; you'll also have to fund our work for the next five or ten years that it will take to achieve this. You will bear 100% of our research and development costs, and in return, we'll share 50% of the profits when the time comes.

This model has allowed us to continue making groundbreaking discoveries. What's remarkable, even unbelievable, is that some of our partners, like Santa Fe, stuck with us and are now reaping the benefits of billion-dollar drugs.

But some partners after five, six, or seven years into the process said ohhh this hasn't delivered yet, so they essentially gave it back to us! We only had maybe two or three years left to figure out how to fund it on our own... and these guys lost out!

So Eylea that sells about $8 billion worldwide, a 100% of it comes to Regeneron. Why? Because first it was shared 50-50 with Proctor and Gamble... they gave up on it after four years! Then it went to Ventus, they gave it back after three or four years and Bayer was willing to fund the remaining couple of years of research... However, because things were looking so promising at that point, they only received 50% of the sales outside of the US, while we retained 100% of the sales within the US.

This is how we funded our business, by getting partners to support our research and development while still granting us a 50% share at the end of the process. Consequently, we now have billion-dollar drugs, some of which we receive 100% of the profits because our partners gave up, and others where we receive 50% of the profits, which is still quite remarkable." — George Yancopoulos, Cofounder of Regeneron

A BIOTECH CEO'S MOST IMPORTANT JOB

"At the heart of strong leadership is the ability to command teams, to build community, to build teamwork, to build communication across executive committees, senior leadership committees, boards... It's the ability to stabilise teams and the communication across teams, up to the board down to the senior leadership team and across the organisation." — Jodie Morrison

Jodie Morrison is a Venture Partner at Atlas Venture and has been a CEO and Board Member of, I don't know, how many biotech companies!

Jodie continues... "I'm always looking for good team players and people who have high EQ because ultimately you can find all the skills but bad EQ and the inability to integrate properly and communication across a team is to me, the biggest killer we run into in biotech companies is a breakdown in that communication, which I think often can be traced back to problems with EQ."

Again... the "ability to stabilise teams"!

Here's an example of this...

A biotech company is for most of its early life an R&D company. The scientific founders, the scientists, the R&D team, and later the CMO, the regulatory people, the clinical people...

These guys spend years pouring blood, sweat, and tears during the early research, pre-clinical and clinical stages. And they do care a lot; they see the company as their baby, and rightfully so.

And one day when the science and drug discovery hopefully succeed, it's then time for the commercial people to run the show... and rightfully so! Without their help, the company can't create/launch/market a product and reach the patients.

But it can be hard for the scientists to hand over their baby, which they have raised with so much love and care for 10+ years to these new strangers! They can be sceptical, they can be unwilling to let go of control or power if you wish... These are all legitimate human feelings!

The CEO here must be a master "team stabiliser", and manage egos, listen to the concerns of everyone, resolve conflict, and help people trust each other and work well together.

And this happens with every transition; a biotech company goes through so many transitions, i.e. private to public, pre-clinical to clinical, clinical to commercial. Transitions are all about transparency, communication and stabilisation!

John Maraganore has said the same thing: "The biggest challenge in this business is not the science, it's not the money... the biggest challenge is people".

It's all about stabilising and aligning people towards the accomplishment of a shared vision.

That's your job. If you do this one thing well, you're done.

And of course.... you have to raise a lot of $$$ so that you can have people to align!

CREATE YOUR PATIENT WALL.

This is the Patient Wall at Agios Pharmaceuticals, adorned with the faces of people that Agios has helped throughout the years.

It serves as a reminder to everyone at the company that their work affects and saves real lives.

I discovered Agios' Patient Wall through Peter Smith's blog post on on the Life Sci VC's website.

Peter and the team at Remix Therapeutics have created their own Patient Wall:

"Charles Kung, who has been at Agios for many years told Alex Harding about the Patient Wall at Agios and how it inspired the company. We loved that idea and decided to add our own electronic version, a big screen TV with photos of family and friends of Remix employees at our entrance that reminds us each day we are showing up to work to help patients. It has become an important part of our culture for new employees to add photos; we get to know them better and remember we are all touched by disease in some way and our work has REAL meaning." — Peter Smith, CEO at Remix Therapeutics

During an interview with the BIOS Community, Chris Gibson said,

"Here's the best advice I've ever gotten. When it's hard, go find a patient. Because when you're sick, all that matters is being healthy. It can reinstill all of the passion, energy and motivation. Keep yourself close to the patient." — Chris Gibson, CEO at Recursion

And everyone should stay close to the patients, not just the founders/CEOs:

"To build passion in the workplace, the former CEO of Novo Nordisk, Lans Sorensen had ALL the employees meet with patients and understand what their lives were like and how the company's products were transformative..."

Everyone in the biotech community, including the person writing this...we're into this because we want our lives to have made an impact on human health and patients.

But because the daily grind can easily consume us, a five-minute conscious and deliberate weekly habit such as reflecting on our why, standing by the patient wall, talking to a patient, concluding a weekly team meeting with a reminder of our shared vision, can give access to hidden reservoirs of energy and motivation in both good and tough days.

LESSONS FROM OUR HENRI TERMEER

I've spent the last couple of days reading everything I could find about Henri Termeer.

Whenever you need a little motivation in your arduous journey, please read the story of this man. It will do your heart good. It will help you understand why you chose this journey...

Henri was the CEO of Genzyme, a company committed to developing life-saving treatments for rare diseases.

When he joined as CEO, the company didn't have a clear plan about what it wanted to do.

Every weekend on a Saturday, sometimes Sunday too, we would meet either in the combat zone on the 50th floor of 75 Newland Street in Boston or in a classroom at MIT and talk about what we could uniquely do...

We couldn't do much with what we had back then but we could do a lot if we let our minds go wild...

We wanted to do important stuff, to make a big difference.

In tremendous amount of good fortune, we focused on Gaucher, a disease that causes fatty deposits to accumulate in certain bones and organs and affected young children.

What's missing in these patients is an enzyme that normally breaks down lipids in the body.

We found that the enzyme was in high concentration in the placenta.

We had a small car which we would use to visit hospitals and collect placentas.

In the evening, I would go home and the sweet smell of placentas would still be in the car.

We would lift the heavy placentas and take them to our office on the 50th floor where we had two centrifuges.

We had to spin the placentas to get rid of the fluid as we needed only the dry placenta tissue.

When we did that, the whole building would vibrate and all 50 floors would stop working!

We first tested the enzyme on Brian, a four-year-old boy.

To watch a four-year-old child who is slowly degenerating before your very eyes is a terrible experience. When you see a boy like that, you want to help him.

Brian was given an infusion and the effect was immediately visible. After only three or four infusions, he felt a lot better. He was more active and felt healthy.

That was the moment that I remember the most in my life, the moment that motivated me the most in the 30 years I spent in Genzyme.

But then, ten years of ridiculous mountains passed...

Many people would say to me, are you kidding me? This can't work.

The reason was that we needed 20,000 placentas, 20,000 kids to be born, to treat one patient for a year!

Well-known Harvard names said to me we couldn't bet the company on one patient.

We ignored them.

It was actually three of us who really believed: one scientist, Brian's mother, Robin, and myself.

Because I saw the kid, I saw the family, I saw the difference!

So, we went out to raise the money...

I personally travelled around the country to talk to nice-meaning people.

I brought Robin to talk in a room full of investors.

The people in that room got the message and started calling their colleagues all around the country.

In the end, we managed to sell an RnD limited partnership worth $10 billion in the next five days.

And this is how we managed to do a small trial on 12 people.

The FDA had 214 patients. I said to them, I can only do 12 because I don't have enough placentas.

Our enzyme worked on all of them. It was a dramatic difference!

But then, we had to overcome the hurdle of the placentas.

We found a plant in France where they extracted plasma from placentas using winepresses.

I told them, I can't build a plant, but I can give you the money to build one.

You need the fluid, I need the dry tissue.

They declared us crazy... but they agreed to partner with us.

It was a very emotional partnership...

In the end, we managed to treat about a thousand kids.

Visualise the logistics: we processed the enzyme from almost 33 per cent of all placentas from birth in the United States.

Million of placentas would find their way to this little town in France...

This is how we built our culture.

Our culture was: Yes, you can do this. And if no one can see it but you, you can still be right..." — Henri Termeer

WHAT DOES IT TAKE TO BE A GOOD LEADER IN A PHARMACEUTICAL COMPANY?

The founder and former CEO of Vertex Pharmaceuticals, Joshua Boger, was once asked: "What does it take to be a good leader in a pharmaceutical company?"

Boger said, "Pharmaceutical discovery and development is the most complicated activity that humans do.

Going to Mars is easy, going to Mars is engineering! I mean, I'm not an engineer but if you give me a hundred billion dollars, I guarantee you I can put you on Mars. There's nothing stopping me except organisational momentum and money.

But if you give me a hundred billion dollars and say, in five years or ten years you have to cure Alzheimer's disease... I'll say... I'll give it a go but I have no idea whether I could do it!

Drug discovery and development is just an insanely complicated activity to do that you need to keep teams of people together for a very long time.

And this is what distinguishes it from most of the high-tech, i.e. I have a high-tech idea, I should be shipping it in 18 months. If I have a pharmaceutical idea I may be shipping it in 18 years!

So I need to keep a team of people together on the discovery and development side for 12 to 15 years on average.

The most important thing is understanding team dynamics and the difference in personalities that are necessary to build a team and hold it together. To do that, you the leader have to be incredibly passionate about what you're trying to do. It's a combination of empathy and passion...

And you have to lead by example. Very often I was the last person to cut the lights out at midnight or 1 am. And it's not just the number of hours... the people who report to you can see you, and they can tell whether or not you're thinking about the mission all the time..."

A COMMON PITFALL FOR FIRST-TIME BIOTECH CEOS

You have probably grown up within a particular function, i.e. clinical, medical, research, business, etc.

And, all of a sudden, as CEO you're responsible for all functions!

Now here's the thing...

You don't need to become an expert overnight in all of those other functions where you didn't grow up.

Because first, you can't, and even if you could, you shouldn't do it anyway!

Because you're not the soloist, you are the conductor!

And I mean, I can empathise with you when you know that and still want to be the soloist!

It's hard to give up on all these things... you care, you want to have peace of mind, you want more control, who doesn't?

But, your power, your biggest leverage is in finding the best talent, trusting them, creating the best culture, and providing a runway for them to do their jobs.

Your job is not to tell them how to play the violin or the cello.

Because there's nothing more demoralizing for the soloist than being told how to play that note...

You are the conductor... you are the visionary, the communicator, the motivator, the strategist.

Now this is good news... because you can take a step back, breathe, relax.

You don't have to dive in and learn everything about everybody's job.

Just watch them. Do they have everything they need?

Talk to them. Listen for their fears, concerns, and challenges.

Coach them. Assume they have all the answers.

And I'm saying this again... I get it when you sometimes want to borrow their violin, it can also be fun for you, right?

That's okay... they can give it to you for a minute but not for the whole concert, okay???

THE PHARMACEUTICAL INDUSTRY IS AN INFORMATION INDUSTRY

The title above is a quote by the famous management guru Peter Drucker.

If this is true, how does someone win in an "information industry":

First of all, what Drucker meant was that the pill you get may cost a few cents to make but its real value is in the years of R&D, the IP, all the failed attempts before it, etc.

Now, if information is the currency in this industry, to me, a biotech team must be good at:

- generating information
- extracting information from others
- sharing information internally
- sharing information externally

1. To generate information, you have to be willing to innovate. To innovate, you have to deal with your fear of failure.

To deal with your fear of failure, you need a strong purpose.

2. To extract information from others (advisors, investors, pharma, partners) you need to develop strong relationships.

And strong relationships are built on trust and integrity.

3. To share information internally—and do it well—the team members have to manage their egos and put the team and the mission first.

4. To share information externally with partners means that you are aware of your limitations (self-awareness) and you acknowledge that you can't realise your vision on your own.

HOW YOU ATTRACT THE BEST TALENT WITHOUT TRYING TOO HARD

Here are some wise words from Amgen's former CEO, Gordon Binder:

"When George Rathmann interviewed me for the position of chief financial officer, he made no attempt to sugarcoat the seriousness of Amgen's looming financial crisis. Nor should he have. The company needed a CFO who would be up to the challenge of obtaining private or public funding, and I needed a vivid picture of what lay ahead so that I could hit the ground running.

Some organizations try too hard to impress. They set their sights on a hot prospect and work overtime selling him on the organization even if it means revealing only selective information to paint a rosy picture. That's a big mistake. A job interview should be a mutual exploration of whether the two parties would make a harmonious match, with full disclosure on both sides.

Do you know the #1 reason that people leave their jobs within six months? It's feeling blindsided by unrealistic expectations, about either the duties of the job itself, the company, or their role. To be sure, it's up to job applicants to ask questions until they're satisfied that they understand what will be expected of them. But some companies make pie-in-the-sky promises—about future promotions, improved facilities and equipment, and so on—that they know aren't likely to happen anytime soon.

Never try to snow applicants. You might succeed and be sorry you did! When the hard realization settles in that your new hire was misled, whether it was a bald-faced lie or acts of omission, you have lost her for good. The sense of betrayal burns too intensely to be extinguished.

Always tell applicants the whole truth. In that way, if you should make the mistake of offering the job to someone who isn't right for the position, hopefully, he'll take himself out of the running sparing you from reaching the same conclusion after the fact. It's like self-selection but in reverse..."

(From the book Science Lessons by Gordon Binder)

THERE'S ALWAYS SOMEONE WILLING TO BET ON YOU...

"The first deal we did was with Merck in 2003. That was a very small *starter* deal, $7.5 million upfront, driven by Steve Friend who had a real passion for this space and the science. These early deals are often driven by the passion of an R&D leader to do something bold. Steve was the visionary at Merck who said... let's do a deal with Alnylam." — John Maraganore, Former CEO of Alnylam

This is the Power of One. You need only ONE who believes in you and your idea.

There is at least one person in big pharma who can see what you have seen—and they're willing to move mountains for you.

Once you find that one believer, everything else becomes irrelevant... the economy, markets, macro environment... they don't matter! (Even IRA!)

Investors? Same thing... When you're fundraising, you're looking for a market of ONE.

There's always ONE irrational and unreasonable investor out there who's willing to bet on you.

And then, they'll go and convince all the partners, and more VC firms will join until ultimately they'll all prove to be rational and reasonable to bet on you...

Regulators? Same thing... You don't get approved by the FDA; you get approved by people.

You get approved by a "Chief of Oncology and Hematology at the FDA who loses his wife to ovarian cancer and declares a jihad to streamline the review process..." (Read For Blood and Money by Nathan Vardi, p.150)

Now, how do you find the ONE?

You keep searching for the one, you become better at listening, you find intersections between what you care about and what they care about, you keep building and maintaining those relationships, you show them how much you care, you take nothing personally, and you keep believing when there doesn't seem to be any believers around.

HOW TO IGNORE THE NOISE AND PLAY THE LONG-TERM GAME

Luke Timmerman asked Daphne Zohar on his podcast the Long Run: "It's been 15 years since you started. Now you have a few things to point to but how did you get through those linear times when capital was not abundant and you couldn't point to products on phase 2/3 that had the scientific community all jazzed up?"

Daphne said, "This is an industry where it takes a long time to move innovations from discovery through approval. And there are all these incentives around short-term things like... Did you raise that round? But these things are not always indicative of actual progress and getting new treatments to patients. For me, it's about trying to not get caught up in those near-term success markers, and focusing on doing the right work, doing good science, getting the right people, running the right experiments, and asking the right questions."

Wonderful advice.

Now, the question is... how do you do that for 15-20 years and how do you manage not only your own expectations of short-term success but everyone else's too, i.e. investors, markets, press, and other stakeholders?

How do you ignore the Sirens and stay focused on reaching your Ithaca?

Everyone talks about having a bold vision that will keep your team motivated, practicing gratitude for how far you've gone, focusing on building and not paying attention to the hype of the news...

That's all great but there's a crucial intermediate step that particuarly very smart and driven CEOs and scientists miss:

Embracing that you may often crave short-term success. Who doesn't? Evolutionarily, you're wired to seek instant gratification.

You want to generate that data, you want to do that deal, you want to show progress to investors; today, not in one, two, three years!

You feel great responsibility on your shoulders, you care about the team, you care about the vision.

And then you're reading the news... "This company has raised its Series B, that company has a development candidate, they're moving something into the clinic, they're on Endpoints News..."

You may feel impatient, afraid, worried, frustrated, angry, or not good enough.

These are all 100% valid feelings! Don't run away from those feelings. Embrace them.

Feeling fear? Breathe in fear. Feeling anger? Breathe in anger.

The cracks in your resilience come from suppressing those feelings.

Incredibly smart and capable CEOs/scientists will come up with 100 rational reasons why they shouldn't be feeling what they're feeling!

Yes, your brilliant brain is right. But your survival/emotional brain doesn't give a s*** whether you're right.

If your survival brain has perceived a threat, it's going to send you fear; it's a done deal.

When you rationalise, you create more friction, more dissonance, more inner war!

All those emotions and anxieties need is to be acknowledged. When you slow down to embrace them, they wash away.

And then, of course you can use your brilliant brain to solve biology and discovery problems. For those problems, your brain is just perfect!

But please don't use your brain to fix your emotions.

Don't try to kill the "worrior", the inner-critic, the impostor.

All these parts of yourself are coming from your deepest spirit that wants you to succeed and look after yourself, the team, and the vision.

This is where the role of a coach can be pivotal.

You need the support of people who can appreciate all parts of you.

You need someone to slow you down from the race and help you bring all these emotions to conscious awareness.

You need a safe space where you can vocalise all your concerns, worries and anxieties—and be loved regardless.

Loving you, Angelos.

START WITH UNREASONABLE BETS, BUT THEN BE NIMBLE, AND FOLLOW THE DATA

Noubar Afeyan says, "Scientific innovation is heavily limited by reasonableness. How can you expect extraordinary results from reasonable people doing reasonable things?

Everybody's looking for extraordinary outcomes, but they constantly want to figure out whether every single step is the right, reasonable one.

People go ask key opinion leaders, they do what's called due diligence and there's this massive crowdsource reasonableness filter. And yet they expect extraordinary results...

I learned a long time ago, that if you aspire to extraordinary results, you've got to be comfortable with being unreasonable—and persistently so.

So you might say, 'Well, then, if you're unreasonable, does that mean you get extraordinary results?' Of course, not; it's not reversible... And that's where the evolutionary approach comes in...

You first project yourself to things and places that people consider unreasonable in terms of what you're proposing to do or are willing to do...

But then what makes you survive is recognizing that when you get there, you can do variation and selection and look for what survives." — Noubar Afeyan, Founder and CEO at Flagship Pioneering

---

John Maraganore quotes George Bernard Shaw every day while mentoring new biotech CEOs, "The reasonable man adapts himself to the world; the unreasonable one persists in trying to adapt the world to himself. Therefore all progress depends on the unreasonable man." He even includes this quote in his email signature!

At the same time, when I asked him recently, what is some advice that you give to CEOs again and again, he replied: "I tell them that they have to be truth seekers. They have to follow the science and what the results are showing them. They have to be objective. Sometimes I've seen leaders who become overly zealous around their science and technology and often delude themselves.

Do the right things, follow the science, but also create some optionality. Maybe the road that you've got put on is not the road that you may have to stay on forever. When you're pioneering a whole new approach/modality, you can see yourself as an explorer in a new world. Look at everything that's there, use all the tools, be curious and observant. Especially in the early days, you have to keep options open, and when you find where you need to go, when you find that opportunity, then you need to focus." — John Maraganore

---

Henri Termeer believed in a similar approach to John's but used a different term to say "explorer".

Henri believed in "being opportunistic", and strategic plans got in the way of that. The more formalized a plan you had, the more it interfered with communication. "You wanted to be nimble and opportunistic", he would say, "not bureaucratic".

Henri hesitated in accepting detailed forecasts beyond 18–24 months. He believed that an overreliance on long-term planning could be paralysing.

When Henri focused on Gaucher disease, he placed a too unreasonable bet, but he had a piece of data that motivated him to pursue it:

"Many people would say to me, are you kidding me? This can't work. The reason was that we needed 20,000 placentas, 20,000 kids to be born, to treat one patient for a year. Well-known Harvard names said to me we couldn't bet the company on one patient. We ignored them. Because I saw the kid, I saw the family, I saw the difference! So, we went out to raise the money..."

The way I see it is... you should never "land yourself" first, you have to let science or life "land you" when you're being too unreasonable in your approach!

But the first default step should always be an unreasonable one. What if? It may sound crazy, but why don't we give this a try?

WHEN TWO PASSIONATE BIOTECH CEOS AND FIERCE COMPETITORS COME TOGETHER AGAIN FUELED BY THEIR LOVE FOR PATIENTS

The narrative revolves around John Maraganore and Stanley Crooke, former CEOs of Alnylam and Ionis Pharmaceuticals.

JM: "When we were preparing for our IPO in early 2004, we were surprised and puzzled by a claim from Ionis stating that our activities related to an siRNA therapeutic infringed certain Ionis chemistry patents. We were in the early stages of R&D and years away from commercializing an siRNA therapeutic and thus covered under the 'safe harbor' from infringement claims afforded drug developers. Also, we had done extensive diligence on the IP landscape for RNAi at founding and believed we could operate outside of any existing IP. Yet we did recognize that the Ionis claim could affect our IPO, so we initiated negotiations with them."

SC: "There were a lot of other siRNA would-be companies at the time and we were entertaining proposals from all of them. When I met John and Barry Greene I said... I don't know whether siRNA is ever going to be a drug, but if it's going to be a drug it will be with these guys—and so we ended up doing that license. And from there we collaborated on many projects including co-founding a company."

JM: "The synergy allowed us to do some really good science together. We had very frequent engagements and we also found ways in which we could avoid competition with each other on specific programs which was a smart thing to do. It made sense from a business standpoint. At the same time, Ionis had a partner in the siRNA space that they could count on and Ionis benefited economically from our success as well."

SC: "So we did all kinds of great stuff together and then... [in 2015] we had a couple of drugs that were directly competitive and we discovered that... we're both really competitive! We had a falling out and we didn't talk to each other for a few years..."

[In 2020 Stan Crooke stepped down as Ionis' CEO to focus on his n-Lorem Foundation which works to provide personalized medicines to the rarest of rare disease patients (nano-rare).]

SC: "As I thought about the patients that we're trying to serve at n-Lorem, I realized that I needed all the help I could get and John could be an incredibly helpful person for these patients. It made no sense to hang on to whatever bitterness I felt. Still I was very worried when I wrote that email to John asking if he'd like to talk."

JM: "I never stopped admiring Stan and I was watching the n-Lorem story from afar and thought it was powerful. There's a phrase in Talmud that 'whoever saves a single life saves the whole world'. We ultimately got back together. To me it reminded me of a big brother and a younger brother—they get into a fight and but they still come together at the end because they are family."

SC: "We did come together around a family, and that family is the nano rare patients. And it's vital to these patients that the two of us work together. I am embarrassed to have waited as long as I did to write to John..."

ps: As I was researching the story, I came across this picture on Stan's Twitter and I emailed John to get more context. John wrote: "Stan and I were friends and competitors at the same time... what my daughters call "frenemies," but mostly friends because of our mutual respect for each other! Stan pioneered antisense technology and our success with RNAi couldn't have happened without him and his passion and zeal. The picture with Henri solidifies our current commitment together to do EVERYTHING for patients, especially for those "forgotten" with N=1 diseases, where most companies won't go. I'm so grateful to know him and have worked with him."

9 TAKEAWAYS FOR BIOTECH CEOS FROM THE BIOTECH HANGOUT AT THE BIO CONFERENCE IN BOSTON (JUNE 2023)

1. REMAIN PRIVATE OR GO PUBLIC?

Private means you don't have a perceived value of your company that's hanging over your head when you're negotiating with potential partners, acquirers or investors.

Also, when you have some new data you can take your time to understand it before you put it out there. As a public company CEO, you have to put all material information out within hours.

One downside of being a private company these days is less access to capital, i.e. more aggressive terms and "down rounds" [when the company accepts a lower post-money valuation than at its previous financing] — Daphne Zohar

Stay private longer to mature your profile and your pipeline, get the right team in place, and get ready to go public when the markets are more robust. — John Maraganore

2. BE VERY OPEN WITH YOUR INVESTORS WHEN RAISING YOUR NEXT ROUND.

Investors are starting to appreciate that it's okay to have a contingency plan. You can go out there and say I want to raise this much but I have a doable plan with a lower amount. — Grace Colon

3. ALWAYS BE EFFICIENT—IN BOTH GOOD AND BAD TIMES.

I've seen companies that overspend when the tide is high. Don't act like money is cheap; it is not. — Daphne Zohar

4. LOOK FOR SOURCES OF NON-DILUTED FUNDING.

We have a person in our team who is focused only on getting grants and non-diluted sources of funding. That gives us a lot of strength to pursue our vision. — Daphne Zohar

5. FORGET THE STOCK MARKET AND FOCUS ON WHAT'S BEST FOR YOUR COMPANY.

Ask the right questions: What stage are we at? When do we have to raise? How much? Chasing the money when times are hot may not be the best individual decision for your company. It may put you in a situation where you can't raise money even when you have things going well. — Brad Loncar

6. LAYOFFS ARE PAINFUL BUT YOU HAVE TO LIVE ANOTHER DAY.

We did two layoffs at Alnylam and they were the most painful thing I've ever had to do as a CEO, but they were necessary. But one thing that we did do was, we kept a list of all the people that had to leave and every week in our management team meeting we would review that list to find out where everybody was. I would also personally call colleagues in the community and say "hey, I have this excellent regulatory person, this research associate..." — John Maraganore

7. SHOULD I FOCUS ON A PRODUCT OR A PORTFOLIO?

It depends on the company. If you've got a disruptive platform company, you have to focus on both the platform and the products that are coming out of that platform. If it's just a company with two products, then yes, this might be a different case. — John Maraganore

8. CAREFULLY DESIGN YOUR COMPANY AND PREPARE FOR BAD TIMES.

If you're always trying to recalibrate your work based on whether financing in the stock market is good or bad, you'll always be playing catch up. You want to be ahead of the curve, not behind the curve. If the right thing for you is to be a platform company, then be a platform company and raise enough money even during bad times to make that happen. If you're constantly changing, that's not a good thing. — Brad Loncar

9. PUBLIC MARKETS ARE BIOTECH'S FACE TO THE WORLD

"It's bad for the whole biotech industry when immature companies go public. Some say that there's no difference between the private and the public markets; they're just different funding sources. I disagree with that. The public market is the public face of our sector to the world.

When investors in the public markets get burned on biotech, that's a bad thing for the whole sector. The fact that the bubble has deflated and the companies that are now going public are good quality companies who are either at or close to the proof of concept stage, it's a good thing for our sector." — Brad Loncar

ps: Brad reminded me of Daphne Koller, the CEO of Insitro, who once said in an interview... "I'll tell you when the public will see that it's serious: When I make medicines that actually help people get better," Koller said. "That's the true validation of this, and going public is frankly just a distraction towards that goal."

Molly He, the CEO of Element Biosciences, shares a similar view: "The commercial maturity, the company maturity, should be there before we say, 'Hey, we're now ready to be a public company. I think this is the piece a lot of companies, small companies rushing to IPO did not see, and I think some are suffering from that today."

GREAT BIOTECH TEAMS HAVE CONFLICT

Exceptional teams have conflict. Because the stakes are high. The people care and are running at full blast. Every conflict is a chance to strengthen and unite the team.

Before we dive in...

You guys are a bunch of stars who:

- are all experts in your fields
- speak different languages
- work in a non-academic environment probably for the first time
- left academia/research because you're hungry for impact

Of course, you'll have conflict! You MUST have conflict.

Conflict means different perspectives come in contact with each other.

A multidisciplinary team's edge is in the conflict itself.

Can you debate openly and fiercely to tackle the complex problem you're called to solve as a team?

This is how you'll win! You're the best in what you do. No one else in this company (or maybe outside the company) has the skills and knowledge you have.

You want recognition, you want glory. That's okay. You deserve it.

But the only way to get all this is by helping the team to win. You're not Google. You're a small company fighting against all odds.

Your personal success is entirely tied to the team's success. You can't afford to prioritise yourself over the team even for a second.

If you're really selfish, you have to put the team first! So how do you have a tough conversation?

How do you speak your truth without ruining the relationship?

Here is my 5-step framework with influences from psychology, neuroscience, nonviolent communication, NLP, and my love for the impact you're here to create:

[STEP #1: ACCEPTANCE]

Name and accept all your feelings. What are you feeling about the other person right now?

Frustration? Anger? Disappointment? Great. All feelings are welcome—and 100% valid. Allow them to be.

Write them down, say them out loud—embrace them.

Smart people use their brilliant brains to "think down" challenging emotions. Not only does this not work, but it also creates more friction!

Let your emotions do their thing and wash away. Use your brilliant brains to figure out the molecule, not to analyse your emotions!

[STEP #2: PREPARATION]

Before the conversation…

2.1 - See things from their perspective.

What information do they have that you may not have? How may your actions have impacted them in a way outside your intentions? What does the situation mean to them?

2.2 - Ensure you have the right setting for the conversation.

Is the physical environment conducive? Do you have privacy? Do you have all the time you need? Are you both in a good state to have a frank discussion? Find a neutral place, a park, or a coffee shop.

[STEP #3: EXPRESSION]

3.1 - Detoxify your message.

No criticising, no judging, no blaming, no evaluating. "You want all the glory for yourself! You always cut me off in the all hands." These statements won't get you what you want; they'll make them reactive. Instead…

3.2 - Focus only on facts.

"You interrupted me twice in the meeting yesterday when I tried to express my concerns about the data. You promised to send a separate email about this to Sam and copy me in but you didn't."

3.3 - Communicate the core feeling in relation to what they said or did.

"I'm worried that we'll waste our time and will have to do everything all over again. I'm afraid that if we miss deadlines, our investors will lose confidence in us."

[Note: I wrote "in relation to" above because, WE are responsible for our own feelings—no one else! If we "hardwire" our feelings to the words or actions of others, we end up having zero agency, i.e. we give full control of our well-being to the other person.]

3.4 - Communicate your core need that is not being met.

"If there's something wrong with the data, we have to address it now. We don't have the time and money to run the experiment all over again. I want to make sure we're on track for our next raise by Q2."

So, what you want to communicate is not "you are rude and disrespectful" but "I am worried about x, I need y".

This is how you help the other person see the impact of their actions—and then clarify, rectify, or help you get what you want.

[STEP #4: LISTENING]

4.1 - Listen for the other person's core needs and feelings.

"This is how I see things. Does it make sense? Tell me how you see things now or if there's something I miss here."

4.2. - If they get into a defensive or blaming mode, help them filter out the toxicity from their message.

"I see you're frustrated, what were you trying to achieve by doing x? What makes you think that there's no reason to worry here?"

4.3 - If they mention that they feel hurt/insulted/criticised, accept it and explain.

"I'm sorry to hear that you feel hurt. It was not my intention to hurt you.

I wanted to communicate my concerns to you about the quality of the data and see what you think."

[STEP #5: ACTIONS]

Final step: Propose concrete actions.

"Can we meet for a couple of hours tomorrow morning to go through all the data before we prepare the report? Can you send an email to Sam to see what he thinks? What else do you need from me?"

Remember that every tough conversation is a hidden opportunity to strengthen a relationship. The more team members follow this process and manage to resolve issues, the more trust they build in one another.

WHY BIOTECH CEOS SHOULD BE OUTSPOKEN ON POLICY MATTERS

"If you're in this industry, you have to be vocal on some of these policy matters. It's very important that CEOs are willing to speak out. I reject this concept that CEOs should just abide to a Milton Friedman approach and stay silent on any social matter. I think that's wrong. That's not how we should operate as a society. And I think we've seen political environments where frankly, the only sane people in the house are from the private sector." — *John Maraganore*

"The thing that keeps me up at night is that policy positions are being put forth without a true diagnostic of the underlying problem. We end up damaging what is a really vibrant industry in our attempt to do good. I think that's what keeps me up at night. That's what worries me about the future. I'm hopeful that it won't happen. I think it requires all of us to get out there, to educate the general public, to educate those in positions of power about what drug development is all about, what it actually takes and how companies work, and how companies need investment dollars to do the next phase of development. It's certainly taken me out of my shell. I've

been more open discussing some of these topics and I think others in the industry have as well." —

Aoife Brennan, CEO at Synlogic Therapeutics

Three meditations on the above:

1. Being outspoken is not about arguing or fighting; it's about communicating, educating, giving policymakers a heads up about how their actions are affecting the future of innovation, the patients, and the wellbeing of an entire industry.

2. What can help a biotech leader communicate effectively and be persuasive is understanding that when policymakers agree on a new policy that happens to negatively affect the whole industry, doesn't necessarily mean that the direct intention behind that policy was to harm the industry. That's the common cause-and-effect cognitive bias.

3. The above ideas don't apply only to conversations with policymakers but any conversation within the company and outside the company, in real life. It's the art of communication. You have more chances of getting what you want when instead of judgding and critisising the other side, you simply communicate to them how their actions are affecting your wellbeing. At the end of the day, they may be totally unaware that their actions are harming you...

WHY YOU SHOULD TALK TO ALL STAKEHOLDERS EARLY ON

"Even in the early days with seed financing, I was having coffee meetings with physicians, I was talking to patients, I was talking to all kinds of people, all kinds of stakeholders to really understand the perspective and be able to better put forth the story, the unmet need, the commercial strategy, and the value proposition for what we were doing.

Even with just coffee meetings and a couple of $1,000 here and there with patient surveys, we were able to put together quite an exciting story that enabled us to raise our seed investment first and then our Series A.

It's very important from the very beginning to spend some time talking to all stakeholders. And that story can evolve, you can hone your story and positioning and perfect it over time.

You need to get that constant feedback and it's not that hard. You can talk to people, you can pick up the phone, you can go on LinkedIn. If you don't have a lot of money, it's amazing the resources and tools that are out there and experts that are willing to talk to you. Once you get a little funding, you can do paid interviews and build on from there. There are survey tools, there's all sorts of ways to get that input.

The important thing is to show that you're going about it the right way, that you're looking at the entire picture, not only as it is today, but as it would be in the future and getting all of that input.

The payer aspect is very important. Very early on, people are going to ask about coverage and about the payer incentive to do this. It's very important to get that as well.

I cannot stress enough that having been an investor as well, those questions need to be answered early. It doesn't have to be a huge quantitative study. You add to that, you refine it as you mature as a company. It can be a handful, but it's actually very powerful to even have those handful of discussions early on." — Grace Colon, Board Member at BIO

YOU CAN'T CLIMB ANNAPURNA WITHOUT A GREAT TEAM

Quote: "In the mountains, the right team is not a luxury; it's a necessity." — Gerlinde Kaltenbrunner, the second woman to have climbed the fourteen mountains that are more than 8,000m

Annapurna is the deadliest mountain with a fatality rate of 32%, far higher than Everest's 14%.

I enjoy watching those Netflix shows about mountain conquests.

What caught my attention is that in each of those shows, people talk about why they picked their teammates.

They talk about their teammates' skills, character, personality, bonding, and trust. They often say, "I decided to do it only because he or she was going to climb with me".

These folks are obsessed with the teams.

Moving a drug beyond the Valley of Death (with more than 90% drug fatality rates) makes Annapurna seem easy to conquer.

This means you, as a biotech CEO/leader, have to be even more obsessed with having the right people on your team than those mountain climbers! Actually, *obsessed* isn't the proper word; I'd better say *neurotic*...

My CEOs often struggle with letting someone go. They say to me, "Maybe I should give them another chance" when it's clear to both of us that they've already let go of this person deep inside.

Giving people a second chance is a great thing but if you've already made up your mind, there is no benefit in doing the uncomfortable thing later.

And I say to them, "Forgive me. It's easy for me to say let them go because it's not me who's going to do it."

Sometimes it's tricky. For example, the person has been there for a while or if you brought them on and it didn't work out, or maybe the company's vision has changed and this person's expertise is no longer needed.

My CEOs can often get really angry against themselves when they realise that they brought in the wrong person only three months after they hire them. And what if that person is a senior executive? And what if you are a public company? And what if it's not the first time that it didn't work out? What do you say to the board? How do you write that press release? The solution can often be accepting that you have made the wrong choice and forgiving yourself for that.

Whatever the solution is, you have to protect the team, the vision, and that person at the end of the day.

Help this person find a new mountain and you go find the right person for your Annapurna.

It's responsibility towards yourself, them, the company, and the patients.

And it's not just about letting someone go; the focus should be on getting the right person in the first place.

Especially with senior leaders... Spend time with them, do the reference checks, do dinners...

And if the right person is on another expedition now, don't give up; go after them!

Sujal Patel at Nautilus Biotechnologies said in an interview that he spent a year recruiting his Chief Business Officer, "I turned my search off because after I met Nick who wouldn't interview with us I was like... that's the guy! It took me one year to get him to join us."

Good luck to all of you guys who are climbing Annapurnas.

ps: This post was inspired by my recent conversation with John Maraganore. What's fascinating to me is that John mentioned the word journey a couple of times when talking about "having the right people with you" and he then said, "One person cannot do it alone. Leaders need someone else to help them close the door they opened, complete the sentence or see around the corner. For me at Alnylam it was Barry Greene and Akshay Vaishnaw who made all the difference."

WHY HE AND
WE LOVE BIOTECH...

When Stelios Papadopoulos, aka the Godfather of biotech, was asked why he loves biotech, he said:

"What intrigues me the most is this whole process of taking a casual idea in the lab. Sometimes you sit with a cup of coffee and you say, gee, how about this? And then, twenty five years later, there's somebody with kidney cancer who responds to a drug...

That only happened because of this unique interplay of scientists, drug developers, clinical scientists, people betting money on stocks, people who build companies, regulators, journalists, writers, and movie makers.

It's a fascinating laboratory that's multidisciplinary and highly complex.

And all of this, in the end, is for something good. Lives are saved, pain is relieved.

I don't think there's anything more fascinating than this in life.

There is fascinating stuff, eg. understanding string theory. It's intellectually, and potentially satisfying, but it's hard to see the immediate positive effect on human life..."

(From Nature Biotechnology's First Rounders Podcast with Brady Huggett)